# Carpentry Workbook

**Timothy Lockley**

Delmar Publishers

I(T)P  An International Thomson Publishing Company

Albany • Bonn • Boston • Cincinnati • Detroit • London • Madrid
Melbourne • Mexico City • New York • Pacific Grove • Paris • San Francisco
Singapore • Tokyo • Toronto • Washington

## NOTICE TO THE READER

Publisher does not warrant or guarantee any of the products described herein or perform any independent analysis in connection with any of the product information contained herein. Publisher does not assume, and expressly disclaims, any obligation to obtain and include information other than that provided to it by the manufacturer.

The reader is expressly warned to consider and adopt all safety precautions that might be indicated by the activities herein and to avoid all potential hazards. By following the instructions contained herein, the reader willingly assumes all risks in connection with such instructions.

The publisher makes no representation or warranties of any kind, including but not limited to, the warranties of fitness for particular purpose or merchantability, nor are any such representations implied with respect to the material set forth herein, and the publisher takes no responsibility with respect to such material. The publisher shall not be liable for any special, consequential or exemplary damages resulting, in whole or in part, from the readers' use of, or reliance upon, this material.

**Delmar Staff**

Publisher: Susan Simpfenderfer
Senior Editor: Mark Huth
Developmental Editor: Jeanne Mesick
Project Editor: Patricia Konczeski
Art/Design Coordinator: Cheri Plasse
Production Coordinator: Dianne Jensis
Marketing Manager: Lisa Reale

COPYRIGHT © 1995
By Delmar Publishers
a division of International Thomson Publishing Inc.

The ITP logo is a trademark under license

Printed in the United States of America

For more information, contact:

Delmar Publishers
3 Columbia Circle, Box 15015
Albany, New York 12212-5015

International Thomson Editores
Campos Eliseos 385, Piso 7
Col Polanco
11560 Mexico D F Mexico

International Thomson Publishing Europe
Berkshire House 168 – 173
High Holborn
London, WC1V 7AA
England

International Thomson Publishing GmbH
Königswinterer Strasse 418
53227 Bonn
Germany

Thomas Nelson Australia
102 Dodds Street
South Melbourne, 3205
Victoria, Australia

International Thomson Publishing Asia
221 Henderson Road
#05 – 10 Henderson Building
Singapore 0315

Nelson Canada
1120 Birchmount Road
Scarborough, Ontario
Canada M1K 5G4

International Thomson Publishing Japan
Hirakawacho Kyowa Building, 3F
2-2-1 Hirakawacho
Chiyoda-ku, Tokyo 102
Japan

All rights reserved. No part of this work covered by the copyright hereon may be reproduced or used in any form or by any means—graphic, electronic, or mechanical, including photocopying, recording, taping, or information storage and retrieval systems—without the written permission of the publisher.

2 3 4 5 6 7 8 9 10 XXX 01 00 99 98 97

Library of Congress Catalog Card Number: 94-8421
ISBN: 0-8273-6701-5

# Contents

| | | |
|---|---|---|
| Chapter 1 | Wood | 1 |
| Chapter 2 | Lumber | 5 |
| Chapter 3 | Rated Plywood and Panels | 9 |
| Chapter 4 | Non-Structural Panels | 13 |
| Chapter 5 | Laminated Veneer Lumber | 15 |
| Chapter 6 | Parallel Strand and Laminated Strand Lumber | 17 |
| Chapter 7 | Wood I-Beams | 19 |
| Chapter 8 | Glued Laminated Beams | 21 |
| Chapter 9 | Nails, Screws, and Bolts | 23 |
| Chapter 10 | Anchors and Adhesives | 25 |
| Chapter 11 | Layout Tools | 29 |
| Chapter 12 | Boring and Cutting Tools | 33 |
| Chapter 13 | Fastening and Dismantling Tools | 37 |
| Chapter 14 | Saws, Drills, and Drivers | 41 |
| Chapter 15 | Planes, Routers, and Sanders | 45 |
| Chapter 16 | Fastening Tools | 49 |
| Chapter 17 | Circular Saw Blades | 51 |
| Chapter 18 | Radial Arm and Miter Saw | 55 |
| Chapter 19 | Table Saws | 57 |
| Chapter 20 | Understanding Drawings | 61 |
| Chapter 21 | Floor Plans | 63 |
| Chapter 22 | Sections and Elevations | 67 |
| Chapter 23 | Plot and Foundation Plans | 71 |
| Chapter 24 | Building Codes and Regulations | 75 |
| Chapter 25 | Leveling and Layout Instruments | 79 |
| Chapter 26 | Laying Out Foundation Lines | 81 |
| Chapter 27 | Characteristics of Concrete | 85 |
| Chapter 28 | Forms for Slabs, Walks, and Driveways | 89 |
| Chapter 29 | Wall and Column Forms | 93 |
| Chapter 30 | Stair Forms | 95 |
| Chapter 31 | Types of Frame Construction | 97 |
| Chapter 32 | Layout and Construction of the Floor Frame | 99 |
| Chapter 33 | Construction to Prevent Termites and Fungi | 103 |
| Chapter 34 | Exterior Wall Frame Parts | 105 |
| Chapter 35 | Framing the Exterior Wall | 109 |
| Chapter 36 | Ceiling Joists and Partitions | 113 |
| Chapter 37 | Backing, Blocking, and Bases | 115 |
| Chapter 38 | Metal Framing | 117 |
| Chapter 39 | Wood, Metal, and Pump Jack Scaffolds | 119 |
| Chapter 40 | Brackets, Horses, and Ladders | 123 |
| Chapter 41 | Roof Types and Terms | 125 |
| Chapter 42 | Gable and Gambrel Roofs | 127 |
| Chapter 43 | Hip Roofs | 129 |
| Chapter 44 | Intersecting Roofs | 133 |

| | | |
|---|---|---|
| Chapter 45 | Shed Roofs, Dormers, and Special Framing Problems | 137 |
| Chapter 46 | Trussed Roofs | 139 |
| Chapter 47 | Stairways and Stairwells | 143 |
| Chapter 48 | Stair Layout and Construction | 147 |
| Chapter 49 | Thermal and Acoustical Insulation | 149 |
| Chapter 50 | Condensation and Ventilation | 151 |
| Chapter 51 | Cornice Terms and Design | 155 |
| Chapter 52 | Building Cornices | 157 |
| Chapter 53 | Gutters and Downspouts | 159 |
| Chapter 54 | Asphalt Shingles | 161 |
| Chapter 55 | Roll Roofing | 165 |
| Chapter 56 | Wood Shingles and Shakes | 167 |
| Chapter 57 | Flashing | 169 |
| Chapter 58 | Window Terms and Types | 173 |
| Chapter 59 | Window Installation and Glazing | 177 |
| Chapter 60 | Door Frame Construction and Installation | 181 |
| Chapter 61 | Door Fitting and Hanging | 183 |
| Chapter 62 | Door Lock Installation | 185 |
| Chapter 63 | Wood Siding Types and Sizes | 187 |
| Chapter 64 | Applying Horizontal and Vertical Wood Siding | 189 |
| Chapter 65 | Wood Shingle and Shake Siding | 191 |
| Chapter 66 | Aluminum and Vinyl Siding | 193 |
| Chapter 67 | Porch and Deck Construction | 197 |
| Chapter 68 | Fence Design and Erection | 199 |
| Chapter 69 | Gypsum Board | 201 |
| Chapter 70 | Single- and Multi-Layer Drywall Application | 203 |
| Chapter 71 | Concealing Fasteners and Joints | 207 |
| Chapter 72 | Types of Wall Paneling | 209 |
| Chapter 73 | Application of Wall Paneling | 211 |
| Chapter 74 | Ceramic Wall Tile | 213 |
| Chapter 75 | Suspended Ceilings | 215 |
| Chapter 76 | Ceiling Tile | 217 |
| Chapter 77 | Description of Interior Doors | 219 |
| Chapter 78 | Installation of Interior Doors and Door Frames | 221 |
| Chapter 79 | Description and Application of Molding | 223 |
| Chapter 80 | Application of Door Casings, Base, and Window Trim | 225 |
| Chapter 81 | Description of Stair Finish | 227 |
| Chapter 82 | Finishing the Stair Body of Open and Closed Staircases | 231 |
| Chapter 83 | Balustrade Installation | 233 |
| Chapter 84 | Description of Wood Finish Floors | 235 |
| Chapter 85 | Laying Wood Finish Floors | 237 |
| Chapter 86 | Underlayment and Resilient Tile | 241 |
| Chapter 87 | Description and Installation of Manufactured Cabinets | 243 |
| Chapter 88 | Custom-Made Cabinets and Countertops | 245 |
| Chapter 89 | Door Types, Construction, and Installation | 249 |
| Chapter 90 | Drawer Construction and Installation | 251 |

# Preface

In order to meet the needs of students with various learning styles and abilities, a well-rounded educational program needs to include a variety of resource material. The use of this workbook will help address this need. It is designed to be used with *Carpentry* by Gaspar Lewis. Each chapter in *Carpentry* has a corresponding chapter of exercises in the workbook. The objective of these exercises is to provide the student with learning reinforcement of the material in the main text.

After reading each chapter in *Carpentry*, the student should also read the related chapter in the workbook. If need be, it is acceptable for the student to refer back to the textbook for the correct answers. The exercises are comprised of multiple choice, completion, matching, identification, and discussion. Several of the discussion questions are open-ended without exact correct or incorrect answers. These questions are intended to stimulate thought and discussion between the students and instructor.

# Acknowledgments

I wish to thank the administration, faculty, and support personnel of the Vo-Tech and Barnes School of George Jr. Republic.

I also wish to extend my gratitude to the following people who provided me with reviews:

Ron Norrie, North Platte, NE
John A. Weeks, Eastern Michigan University, Ypsilanti, MI
R. Bruce Purdy, Grove City High School, Grove City, OH
Terry Smith, Dorchester High School, Dorchester, NE

# Dedication

This book is dedicated to the young men of George Jr. Republic who have the courage and fortitude to make positive changes in their lives.

Name _____   Date _____

# CHAPTER 1    WOOD

## Multiple Choice

Write the letter for the correct answer on the line next to the number of the sentence.

_____ 1. The carpenter must understand the nature and characteristics of wood to _____ .
A. protect it from decay
B. select it for the appropriate use
C. work it with the proper tools
D. A, B, and C

__A__ 2. The insulating value of 1" of wood is the equivalent of _____ of brick.
A. 6"
B. 10"
C. 14"
D. 18"

__D__ 3. _____ is a wood that is known for its elasticity.
A. Oak
B. Maple
C. Pine
D. Hickory

__C__ 4. The natural substance that holds wood's many hollow cells together is called _____ .
A. pith
B. cambium layer
C. lignin
D. sapwood

__C__ 5. Tree growth takes place in the _____ .
A. heartwood
B. medullary rays
C. pith
D. cambium layer

__B__ 6. The central part of the tree that is usually darker in color is called the _____ .
A. sapwood
B. heartwood
C. springwood
D. medullary rays

__A__ 7. Wood growth that is rapid and takes place in the _____ is usually light in color and rather porous.
A. spring
B. summer
C. fall
D. winter

1

__A__ 8. Periods of fast or slow growth can be determined by _____ of the tree.
  A. counting the annual rings
  B. measuring the height
  C. studying the width of the annual rings
  D. measuring the circumference

__C__ 9. _____ is an example of a hardwood that is softer than some softwoods.
  A. Basswood
  B. Oak
  C. Redwood
  D. Cherry

__A__ 10. All softwoods are _____ .
  A. close-grained
  B. cone-bearing
  C. open-grained
  D. A and B

## Completion

Complete each sentence by inserting the correct answer on the line near the number.

_____ 1. _____ trees lose their leaves once a year.

_____ 2. Softwoods come from _____ trees, commonly known as evergreens.

_____ 3. Water passes upward through the tree in the _____ .

_____ 4. Wood that comes from deciduous trees is classified as _____ .

_____ 5. Fir comes from the _____ classification of wood.

_____ 6. Oak is an example _____ -grained wood.

_____ 7. The _____ of cedar, cypress, and redwood are extremely resistant to decay.

_____ 8. Open-grained lumber has large _____ that show tiny openings or pores in the surface.

_____ 9. Cedar can always be identified by its characteristic _____ .

_____ 10. The best way to learn the different types of wood is by _____ with them.

# Identification: Cross-section of Wood

Identify each term, and write the letter of the correct answer on the line next to each number.

__C__ 1. pith

__A__ 2. sapwood

__D__ 3. cambium layer

__E__ 4. medullary rays

__F__ 5. heartwood

_____ 6. annular rings

__G__ 7. bark

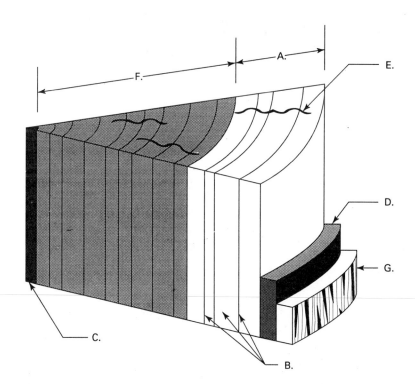

Name _____    Date _____

# CHAPTER 2    LUMBER

## Multiple Choice

Write the letter for the correct answer on the line next to the number of the sentence.

__C__ 1. The process of restacking lumber in a way that allows air to circulate between the pieces is known as _____.
   A. blocking
   B. spacing
   C. sticking
   D. airing

__A__ 2. The best appearing side of a piece of lumber is its _____ side.
   A. face
   B. visage
   C. veneer
   D. select

__B__ 3. Most logs are sawed using the _____ method.
   A. plain-sawed
   B. quarter-sawed
   C. edge-grained
   D. a combination of the plain and quarter-sawed

__B__ 4. Cracked ceilings, sticking doors, squeaking floors and many other problems can occur from using _____ lumber.
   A. recycled
   B. green
   C. seasoned
   D. quarter-sawed

__A__ 5. The moisture content of lumber is expressed as a percentage _____.
   A. of its total weight
   B. of its total volume
   C. of the weight of its free water
   D. of weight to volume

__A__ 6. Wood has reached its _____ when all of the free water is gone.
   A. equilibrium moisture content
   B. stabilization point
   C. fiber-saturation point
   D. dehydration point

__B__ 7. Lumber that is under 2" thick has the classification of _____.
   A. timbers
   B. boards
   C. dimensional
   D. joists

___A___ 8. Dimensional lumber is in the following category: _____
  A. under 2" thick
  B. 2"– 4" thick
  C. 5" and thicker
  D. open-grained only

___B___ 9. The best grade of hardwood as established by the National Hardwood Association is _____.
  A. select
  B. first and seconds
  C. No. 1 commons
  D. choice

___C___ 10. Parallel cracks between the annual rings in wood that are sometimes caused by storm damage are known as _____.
  A. shakes
  B. crooks
  C. checks
  D. cups

## Completion

Complete each sentence by inserting the correct answer on the line near the number.

___Plain___ 1. _____ -sawed lumber is the least expensive method of sawing.

___Quarter___ 2. _____ -sawed lumber is less likely to warp or shrink.

___Combination___ 3. The _____ uses a great amount of skill in determining the most efficient and conservative way to cut a log.

___green___ 4. When lumber is first cut from the log it is called _____ lumber.

___Water___ 5. The heavy weight of green lumber is due to its high _____ content.

___fungi___ 6. The low form of plant life that causes wood to decay is known as _____.

___20___ 7. Wood with a moisture content of below _____ percent will not decay.

___10-12___ 8. Lumber used for framing should not have a moisture content over _____ percent.

___8-10___ 9. Lumber used for interior finish should not have a moisture content over _____ percent.

_____ 10. _____ moisture content occurs when the moisture content of the lumber is the same as the surrounding air.

___4___ 11. S4S means the lumber was surfaced on _____ sides.

Lumber Defects 12. Crooks, bows, twists, and cups are classified as _____.

_____ 13. Four boards 1″ × 6″ × 12′ contain _____ board feet.

_____ 14. Twenty-four boards 2″ × 4″ × 16′ contain _____ board feet.

## Identification: Cut Lumber

**Identify each term, and write the letter of the correct answer on the line next to each number.**

__F__ 1. crook

__A__ 2. quarter-sawed

__D__ 3. twist

__E__ 4. cup

_____ 5. check

__C__ 6. bow

__G__ 7. plain-sawed

A.

B.

C.

D.

E.

F.

G.

## Discussion

**Write your answer on the lines below each instruction.**

1. Describe the difference between air dried and kiln dried lumber.

_____

_____

_____

2. Describe the difference between nominal and actual dimensions.

   _____

   _____

   _____

3. Describe some of the factors one must keep in mind when properly storing lumber on the job site.

   _____

   _____

   _____

Name _____     Date _____

# CHAPTER 3    RATED PLYWOOD AND PANELS

## Multiple Choice

Write the letter for the correct answer on the line next to the number of the sentence.

_____  1. A/an _____ is a very thin layer of wood.
      A. underlayment
      B. com-ply
      C. span
      D. veneer

__C__  2. With the use of engineered panels _____ .
      A. construction progresses faster
      B. more surface protection is provided than with solid lumber
      C. lumber resources are more efficiently used
      D. A, B, and C

__D__  3. Cross-graining in the manufacture of plywood refers to _____ .
      A. touch sanding the grain
      B. the use of open-grained hardwoods
      C. the grain of each succesive layer is at a right angle to the next one
      D. the placement of the peeler log on the lath

__D__  4. The American Plywood Association is concerned with quality supervision and testing of _____ .
      A. waferboards
      B. composites
      C. oriented strand board
      D. A, B, and C

__C__  5. The letters N, A, B, C, and D indicate _____ .
      A. span rating
      B. exposure durability classification
      C. the quality of the panel veneers
      D. strength grades

__B__  6. Douglas fir and southern pine are classified in the _____ strength grade.
      A. plugged C
      B. group 1
      C. 303
      D. 32/16

_____  7. A performance rated panel meets the requirements of the _____ .
      A. panel's end use
      B. sawyer
      C. EPA
      D. U.S. Forest Service

__A__ 8. The left-hand number in a span rating denotes the maximum recommended support spacing when the panel is used for _____.
A. roof sheathing
B. subflooring
C. siding
D. underlayment

## Completion

**Complete each sentence by inserting the correct answer on the line near the number.**

_____ 1. The _____ is the largest trade association that tests the quality of plywood and other engineered panels.

_____ 2. The sheets of veneer that are bonded together to form plywood are also known as _____ .

_____ 3. Plywood always contains a/an _____ number of layers.

_____ 4. Specially selected logs mounted on a huge lath are known as _____ logs.

_____ 5. The highest appearance quality of a panel veneer is designated by the letter _____ .

_____ 6. Panels with a _____ grade or better are always sanded smooth.

__Siding__ 7. V-groove, channel groove, striated, brushed, and rough-sawed, are all special surfaces used in the manufacture of _____ .

_____ 8. Most panels manufactured with oriented strands or wafers are known as _____ .

_____ 9. Composite panels rated by the American Plywood Association are known as _____ .

_____ 10. Exposure durability of a panel is located on the _____ .

## Matching

Write the letter for the correct answer on the line near the number to which it corresponds.

_____ 1. composite panels

_____ 2. span rating

_____ 3. exposure 1

_____ 4. exposure 2

_____ 5. exterior

_____ 6. plywood

_____ 7. aspenite

_____ 8. grade stamp

A. may be exposed to weather during moderate delays

B. is a specific brand of oriented strand board

C. cross-laminated, layered plies glued and bonded under pressure

D. may be exposed to weather during long delays

E. wood veneer bonded to both sides of reconstituted wood panels

F. appears as two numbers separated by a /

G. may be permanently exposed to weather or moisture

H. assures the product has met quality and performance requirements

## Identification: Label Information

Identify each term, and write the letter of the correct answer on the line next to each number.

__D__ 1. thickness

__E__ 2. mill number

__A__ 3. panel grade

__F__ 4. national research board report number

__C__ 5. exposure durability classification

__B__ 6. span rating

```
              APA
A. ─────→ RATED SHEATHING
B. ─────→ 32/16  1/2 INCH ←───── D.
           SIZED FOR SPACING
C. ─────→ EXPOSURE 2
              000 ←───────────── E.
            NRB-108 ←──────────── F.
```

Name _____  Date _____

# CHAPTER 4  NON-STRUCTURAL PANELS

## Multiple Choice

Write the letter for the correct answer on the line next to the number of the sentence.

_____ 1. Non-structural particleboard is used in the construction industry for _____.
  A. underlayment for finish floors
  B. kitchen countertops
  C. the core of veneer doors
  D. A, B, and C

_____ 2. The highest quality particleboard _____.
  A. contains the same size particles throughout
  B. is 100% sawdust
  C. has large wood flakes in the center with the particle size decreasing the closer to the surface
  D. usually has a rough surface texture

_____ 3. High density fiberboards are called _____.
  A. particleboard
  B. softboard
  C. oriented strand board
  D. hardboard

_____ 4. *Masonite* is a trademark for _____.
  A. softboard
  B. duraflake
  C. hardboard
  D. particleboard

_____ 5. Hardboard is widely used for _____.
  A. sound control purposes
  B. exterior siding and interior wall paneling
  C. roof sheathing
  D. insulation

_____ 6. _____ is a brand name for softboard.
  A. *Celotex*
  B. *Fibrepine*
  C. *Masonite*
  D. A, B, and C

_____ 7. To protect exterior softboard wall sheathing from moisture during construction it is impregnated with _____.
  A. lignin
  B. asphalt
  C. oil
  D. ceresote

_____  8. _____ is a well-known reference on products used in the construction industries.
   A. The Dodge Report
   B. Dunn and Bradstreet
   C. Sweets Architectural File
   D. Thompson's Products

## Completion

**Complete each sentence by inserting the correct answer on the line near the number.**

_____  1. _____ plywood is frequently used on the interior of buildings for wall paneling and cabinets.

_____  2. The quality of _____ is indicated by its density per cubic foot.

_____  3. _____ are manufactured as high density, medium density, and low density boards.

_____  4. Lignin is utilized in the manufacturing process of _____.

_____  5. Tempered hardboard panels are coated with _____ and baked.

_____  6. Most hardboard producers belong to the _____ _____.

_____  7. *Medite, Baraboard,* and *Fibrepine* are trade names for _____.

_____  8. Decorative ceiling panels for suspended ceilings are extensively made from _____.

## Math

**Add the following numbers.**

| | 1) | 2) | 3) | 4) | 5) | 6) | 7) |
|---|---|---|---|---|---|---|---|
| | 23 | 54 | 89 | 91 | 11 | 74 | 22 |
| | 55 | 77 | 31 | 82 | 14 | 42 | 61 |
| | 9 | 45 | 81 | 8 | 46 | 98 | 51 |
| | 27 | 65 | 58 | 94 | 32 | 21 | 53 |

| | 8) | 9) | 10) | 11) | 12) | 13) | 14) |
|---|---|---|---|---|---|---|---|
| | 74 | 81 | 93 | 32 | 37 | 80 | 22 |
| | 26 | 47 | 92 | 18 | 21 | 27 | 32 |
| | 56 | 55 | 11 | 19 | 67 | 29 | 56 |
| | 81 | 97 | 45 | 33 | 72 | 85 | 11 |

Name _____  Date _____

# CHAPTER 5  LAMINATED VENEER LUMBER

## Multiple Choice

**Write the letter for the correct answer on the line next to the number of the sentence.**

_____ 1. During the manufacture of laminated veneer lumber, the grain in each layer of veneer is placed _____ to the previous one.
   A. parallel
   B. at right angles
   C. diagonally
   D. at 22.5°

_____ 2. Laminated veneer lumber was first used to make _____ .
   A. automobile frames
   B. airplane propellers
   C. underlayment
   D. canoes

_____ 3. The first commercially produced laminated veneer lumber for building construction was patented as _____ .
   A. *Micro-Lam*
   B. *Gang-Lam*
   C. *Struclam*
   D. *Versa-Lam*

_____ 4. Laminated veneer lumber uses _____ in its construction.
   A. basswood
   B. poplar and aspen
   C. Douglas fir or southern pine
   D. mostly hardwoods

_____ 5. Laminated veneer lumber is widely used in _____ .
   A. wood frame construction
   B. subflooring
   C. exterior siding
   D. interior paneling

_____ 6. A typical 1¾ inch beam of laminated veneer lumber contains _____ layers of veneer.
   A. 3–5
   B. 5–8
   C. 10–15
   D. 15–20

_____ 7. The thickness of LVL veneer ranges from _____ of an inch.
   A. 1/32–1/8
   B. 1/10–3/16
   C. 1/4–3/8
   D. 5/16–7/16

_____ 8. Laminated veneer lumber is suited for _____ .
   A. load carrying beams over window and door openings
   B. scaffold planks
   C. concrete forming
   D. A, B, and C

_____ 9. During its manufacture, laminated veneer lumbers' veneers are _____ .
   A. expanded
   B. densified
   C. coated with oil
   D. bonded with an interior adhesive

_____ 10. The usual thickness of LVL is _____ .
   A. ¾"
   B. 1"
   C. 2" and 4"
   D. 1½" and 1¾"

## Math

**Add the following numbers.**

| | 1) | 2) | 3) | 4) | 5) | 6) | 7) |
|---|---|---|---|---|---|---|---|
| | 22 | 63 | 82 | 84 | 31 | 66 | 11 |
| | 74 | 49 | 95 | 32 | 39 | 21 | 8 |
| | 48 | 50 | 37 | 71 | 19 | 46 | 23 |
| | 41 | 92 | 28 | 62 | 92 | 65 | 16 |
| | 7 | 44 | 3 | 51 | 73 | 38 | 31 |

| | 8) | 9) | 10) | 11) | 12) | 13) | 14) |
|---|---|---|---|---|---|---|---|
| | 67 | 52 | 29 | 90 | 31 | 73 | 71 |
| | 4 | 64 | 32 | 17 | 73 | 51 | 52 |
| | 83 | 31 | 78 | 92 | 11 | 13 | 84 |
| | 45 | 5 | 12 | 94 | 78 | 42 | 31 |
| | 89 | 91 | 39 | 73 | 25 | 28 | 52 |
| | 73 | 24 | 31 | 52 | 37 | 25 | 82 |

Name _____     Date _____

# CHAPTER 6  PARALLEL STRAND AND LAMINATED STRAND LUMBER

## Multiple Choice

Write the letter for the correct answer on the line next to the number of the sentence.

_____ 1. Parallel strand lumber provides the building industry with _____.
   A. sheet materials
   B. small dimension lumber
   C. large dimension lumber
   D. A, B, and C

_____ 2. Parallel strand lumber meets an environmental concern by using _____.
   A. strands not made of wood
   B. small diameter second growth trees
   C. only old-growth trees
   D. mostly hardwoods

_____ 3. Parallel strand lumber is manufactured _____.
   A. using a process that is identical to plywood
   B. much the same as particleboard
   C. using a microwave pressing process
   D. using ultrasound to bond the strands

_____ 4. Parallel strand lumber is a material _____.
   A. that is available in 4' × 8' sheets
   B. that sometimes contains defects like knots and shakes
   C. that can be used wherever there is a need for a large beam or post
   D. whose demand will decrease in the future

_____ 5. In comparison to solid lumber, parallel strand lumber _____.
   A. is consistent in strength throughout its length
   B. has few differences
   C. lengths are not as long
   D. may contain checks

_____ 6. The registered brand name for laminated strand lumber is _____.
   A. *Lam-stran*
   B. *Parallam*
   C. *TimberStrand*
   D. *Struclam*

_____ 7. Laminated strand lumber presently is made from _____.
   A. Douglas fir only
   B. longer strands than those used in parallel strand lumber
   C. surplus over-mature aspen trees
   D. southern pine

_____ 8. Laminated strand lumber _____ .
   A. is made of wood strands under 12" long
   B. is not designed to carry heavy loads
   C. can be made from very small logs
   D. A, B, and C

## Math

**Subtract the following numbers.**

1)  542
   −289

2)  752
   −697

3)  812
   − 34

4)  720
   −436

5)  835
   −749

6)  639
   −621

7)  982
   −326

8)  822
   −599

9)  278
   − 89

10) 298
    −129

11) 8229
    −6545

12) 426
    −239

13) 619
    −250

14) 812
    −301

15) 121
    − 72

16) 921
    − 39

17) 273
    −194

18) 721
    −653

19) 103
    − 59

20) 611
    −282

Name _____   Date _____

# CHAPTER 7    WOOD I-BEAMS

## Completion

Complete each sentence by inserting the correct answer on the line near the number.

_____  1. The "I" shape utilized in a wood I-beam increases its _____ .

_____  2. The _____ of a wood I-beam may be made from laminated veneer lumber or specially selected finger-jointed solid wood lumber.

_____  3. Plywood, laminated veneer lumber, or oriented strand board may be used in the _____ of the beam.

_____  4. The manufacturing process of wood I-beams consists of _____ top and bottom flanges to a web.

_____  5. Wood I-beams are produced to approximate _____ moisture content.

_____  6. Wood I-beams are available in depths from _____ to _____ inches.

_____  7. Beams with larger webs and flanges are designed to carry _____ loads.

_____  8. Wood I-beams are available in lengths of up to _____ feet long.

## Math

Multiply the following numbers.

1) 529 × 29        2) 738 × 77        3) 627 × 52        4) 982 × 72        5) 901 × 68

6) 833 × 72        7) 436 × 639       8) 889 × 29        9) 256 × 34        10) 449 × 22

11) 739 × 592      12) 812 × 323      13) 822 × 859      14) 162 × 451      15) 449 × 724

19

16)  374        17)  176        18)  525        19)  866        20)  929
    ×771            ×725            ×828            ×377            ×499

21)  303        22)  574        23)  829        24)  293        25)  117
    × 45            × 91            ×782            ×236            × 33

Name _____   Date _____

# CHAPTER 8    GLUED LAMINATED BEAMS

## Completion

Complete each sentence by inserting the correct answer on the line near the number.

_____ 1. Glued laminated lumber is commonly called _____.

_____ 2. Glued laminated beams and joists are _____ than natural wood of the same size.

_____ 3. Aside from structural strength, glued laminated beams are _____ as well.

_____ 4. The individual pieces of stock in glued laminated lumber are known as _____.

_____ 5. When a load is imposed on a glulam beam that is supported on both ends, the topmost lams are said to be in _____.

_____ 6. _____ is a force applied to a member that tends to increase its length.

_____ 7. The lower stressed sections of glued laminated lumber are located in the beam's _____.

_____ 8. The sequence of lam grades from the bottom to top of glued laminated lumber is referred to as _____.

_____ 9. Glued laminated beams come with one edge stamped _____.

_____ 10. Glued laminated lumber is manufactured in three different grades for _____.

____ ____ 11. Glued laminated beams are usually available in lengths from ____ to ____ feet in 2 foot increments.

____ ____ 12. The widths of glued laminated beams range from ____ to ____ inches.

21

## Discussion

Write your answer on the lines below each instruction.

1. What are some of the reasons engineered lumber products will surpass that of solid lumber in the future?

   _____

   _____

   _____

2. Over time, will the engineered lumber products industry help or harm the lumber industry?

   _____

   _____

   _____

## Math

Divide the following numbers and round off each answer.

1) 452 ÷ 7 _____
2) 763 ÷ 52 _____
3) 525 ÷ 9 _____
4) 1,868 ÷ 11 _____
5) 778 ÷ 11 _____
6) 663 ÷ 5 _____
7) 1,586 ÷ 80 _____
8) 80,549 ÷ 333 _____
9) 12,732 ÷ 1.65 _____
10) 978.9 ÷ 4.3 _____

Name _____   Date _____

# CHAPTER 9    NAILS, SCREWS, AND BOLTS

## Completion

**Complete each sentence by inserting the correct answer on the line near the number.**

_____    1. Steel nails that are uncoated are called _____ nails.

_____    2. Steel nails that are coated with zinc to prevent rusting are called _____ .

_____    3. _____ galvanized nails have a heavier coating than electroplated galvanized.

_____    4. Moisture reacting with two different types of metal causes _____ which, in time, results in disintegration of the softer metal.

_____    5. The diameter or thickness of a nail is referred to as its _____ .

_____    6. Wedge-shaped nails that are stamped from thin sheets of metal are known as _____ nails.

_____    7. The preferred nail for fastening exterior finish is the _____ nail.

_____    8. Small finishing nails sized according to gauge and length in inches are known as _____ .

_____    9. To prevent them from bending, masonry nails are made from _____ steel.

_____   10. The pointed end of the screw is called the _____ point.

_____   11. Steel screws that have no coating are known as _____ screws.

_____   12. _____ screws are larger than wood screws and are turned with a wrench instead of a screwdriver.

_____   13. The square section under the oval head of a carriage bolt prevents the bolt from _____ as the nut is being turned.

_____   14. The proper tool used to turn a stove bolt is a _____ .

# Identification: Nails, Screws, and Bolts

Identify each term, and write the letter of the correct answer on the line next to each number.

__F__ 1. duplex nail

__R__ 2. finish nail

_____ 3. casing nail

__D__ 4. roofing nail

__F__ 5. common nail

__C__ 6. gimlet point

__B__ 7. threads

__A__ 8. shank

__J__ 9. lag screw

__N__ 10. carriage bolt

__M__ 11. machine bolt

__L__ 12. stove bolt

__H__ 13. staple

__I__ 14. corrugated fastener

24

Name _____     Date _____

# CHAPTER 10   ANCHORS AND ADHESIVES

## Multiple Choice

Write the letter for the correct answer on the line next to the number of the sentence.

_____ 1. A _____ is a heavy duty anchor.
      A. wedge anchor
      B. coil anchor
      C. nylon nail anchor
      D. concrete screw

_____ 2. Holes may be drilled directly in the masonry through the mounting holes of the fixture being installed if the _____ is used.
      A. toggle bolt
      B. drop-in anchor
      C. stud anchor
      D. conical screw

_____ 3. The concrete screw is _____ type of anchor.
      A. a heavy duty
      B. a medium duty
      C. a light duty
      D. available in all three classifications

_____ 4. Lead and plastic anchors are also called _____ .
      A. split-fast anchors
      B. lag shields
      C. hollow wall fasteners
      D. inserts

_____ 5. When using chemical anchoring systems, it is important to _____ .
      A. thoroughly clear the hole of all dust
      B. properly torque the bolt into the system
      C. immediately stress test the bond
      D. A, B, and C.

_____ 6. A disadvantage of using toggle bolts is _____ .
      A. their use is limited to solid walls
      B. if removed the toggle falls off inside the wall
      C. the diameter of the hole has to be the same as the bolt
      D. hole depth is critical

_____ 7. A common name for a hollow wall expansion anchor is _____ .
      A. universal plug
      B. conical screw
      C. self-drilling anchor
      D. molly screw

_____ 8. Conical screws are used on _____.
   A. gypsum board
   B. cement block
   C. strand board
   D. A and C

_____ 9. Joist hangers are a form of _____.
   A. universal anchor
   B. wood to wood connector
   C. hollow wall connector
   D. wood to concrete connector

_____ 10. Polyvinyl acetate is a _____ glue.
   A. yellow
   B. mastic
   C. contact
   D. white

_____ 11. Contact cement is widely used for _____.
   A. applying plastic laminates on countertops
   B. interior trim
   C. exterior finish
   D. framing

_____ 12. _____ is a moisture-resistant glue.
   A. Yellow
   B. White
   C. Plastic resin
   D. Aliphatic resin

_____ 13. _____ is a type of mastic that may be used in cold weather, even on wet or frozen wood.
   A. Urea resin
   B. Resorcinol resin
   C. Contact cement
   D. Construction adhesive

_____ 14. When applying troweled mastics it is important to _____.
   A. apply heavily
   B. be sure the depth and spacing of trowel's notches are correct
   C. thoroughly brush as well as trowel the mastic
   D. mix the proper ratio of hardener to the mastic

# Identification: Anchors, Bolt, and Screw

**Identify each term, and write the letter of the correct answer on the line next to each number.**

_____ 1. self-drilling anchor

_____ 2. stud anchor

__I__ 3. sleeve anchor

_____ 4. drop-in anchor

_____ 5. coil anchor

_____ 6. nylon nail anchor

_____ 7. toggle bolt

_____ 8. expansion anchor

__A__ 9. conical screw

A.

B.

C.

D.

E.

F.

G.

H.

I.

Name _____   Date _____

# CHAPTER 11    LAYOUT TOOLS

## Multiple Choice

**Write the letter for the correct answer on the line next to the number of the sentence.**

_____ 1. Early in carpenter training it is important that _____ is mastered.
   A. the essex board foot table
   B. quick and accurate measuring
   C. finger gauging
   D. the octagon scale

_____ 2. Most of the rules and tapes used by the carpenter have clearly marked increments of _____.
   A. yards
   B. 32nds of an inch
   C. 16 inches
   D. metric conversions

_____ 3. Pocket tapes are available in _____ lengths.
   A. 50 and 100 foot
   B. 35 and 50 foot
   C. 6 to 25 foot
   D. A, B, C, and D

_____ 4. A helpful and accurate device for laying out stock that has to be fitted between two surfaces is a _____ .
   A. slipstick
   B. framing square
   C. speed square
   D. sliding T-bevel

_____ 5. The combination square functions as _____ .
   A. a depth gauge
   B. a layout or test device for 90° and 45° angles
   C. a marking gauge
   D. A, B, and C

_____ 6. The layout tool that can double as a guide for a portable power saw is a _____ .
   A. speed square
   B. combination square
   C. trammel point
   D. framing square

_____ 7. The side of the framing square known as the face is the one that has the _____ on it.
   A. essex board foot table
   B. the brace table
   C. the hundredths scale
   D. the manufacturer's name stamped

29

_____ 8. The most used table on the framing square is the _____.
   A. rafter table
   B. brace table
   C. essex board foot table
   D. hundredths scale

_____ 9. The _____ is used to lay out or test angles other than those laid out with squares.
   A. trammel point
   B. butt gauge
   C. slipstick
   D. sliding T-bevel

_____ 10. In the absence of trammel points, the same type of layout can be made with _____.
   A. a plumb bob
   B. a thin strip of wood with a brad through it for a center point
   C. butt markers
   D. a line level

## Completion

Complete each sentence by inserting the correct answer on the line near the number.

_____ 1. The longer of the two legs of a framing square is known as the _____.

_____ 2. In construction the term level is used to indicate that which is _____.

_____ 3. The term _____ is used to mean the same as vertical.

_____ 4. The _____ is used to test both level and plumb surfaces.

_____ 5. The pair of tubes located in the center of the level are used to determine _____.

_____ 6. When using the line level, it is important that the level is placed as close to the _____ of the line as possible.

_____ 7. Although very accurate indoors, the _____ can be difficult to use outside when the wind is blowing.

_____ 8. The spool and awl were widely used before the _____ became popular.

_____ 9. _____ is the technique of laying out stock to fit against an irregular surface.

## Matching

**Write the letter for the correct answer on the line near the number to which it corresponds.**

_____ 1. folding rule

_____ 2. pocket tape

_____ 3. framing square

_____ 4. plumb bob

_____ 5. chalk line

_____ 6. butt marker

_____ 7. line level

_____ 8. wing dividers

A. gives only an approximate levelness; is not accurate

B. practically useless when wet

C. joints must be oiled occasionally

D. used to mark hinge gains

E. is often called a scriber

F. hook slides back and forth slightly

G. is suspended from a line

H. outside corner is called the heel

## Measuring

**Refer to Figure 11.1 to complete the following sentences. Write your answer on the line near the number.**

_____ 1. The length of line A is _____ inches.

_____ 2. The length of line B is _____ inches.

_____ 3. The length of line C is _____ inches.

_____ 4. The length of line D is _____ inches.

_____ 5. The length of line E is _____ inches.

_____ 6. The length of line F is _____ inches.

_____ 7. The length of line G is _____ inches.

_____ 8. The length of line H is _____ inches.

_____ 9. The length of line I is _____ inches.

_____ 10. The length of line J is _____ inches.

_____ 11. The length of line K is _____ inches.

_____ 12. The length of line L is _____ inches.

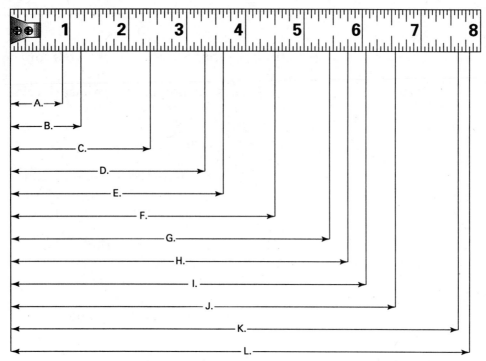

**Figure 11.1**

Name _____ Date _____

# CHAPTER 12 BORING AND CUTTING TOOLS

## Multiple Choice

Write the letter for the correct answer on the line next to the number of the sentence.

_____ 1. Firmer chisels are used mostly on _____ .
   A. finish work
   B. millwork
   C. heavy framing
   D. locksets

_____ 2. The longest bench plane is a _____ plane.
   A. jointer
   B. jack
   C. smooth
   D. block

_____ 3. A bevel that is referred to as hollow ground is one with a surface that is _____ .
   A. convex
   B. concave
   C. flat
   D. raised

_____ 4. When reshaping the bevel on a wood chisel by grinding, it is important to _____ .
   A. use safety goggles
   B. cool the blade by dipping it in water frequently
   C. maintain a width to the bevel that is approximately twice the thickness of the blade
   D. A, B, and C

_____ 5. Chisels and plane irons need to be ground _____ .
   A. every time they become dull
   B. before each time they are used
   C. when the bevel has lost its hollow ground shape
   D. to reestablish a new wire-edge on the tool's back

_____ 6. The handsaw that is designed to cut with the grain is the _____ .
   A. crosscut saw
   B. compass saw
   C. hacksaw
   D. ripsaw

_____ 7. When using the crosscut saw, the saw should be held _____ .
   A. at about 60° to the stock
   B. at about 30° to the stock
   C. at about 45° to the stock
   D. the same angle to the stock as the ripsaw

33

_____ 8. Stock is handsawn with the face side up because _____.
   A. the action of the saw will splinter the bottom side
   B. it prevents the saw from jumping at the start of the cut
   C. it keeps the saw from binding in the cut
   D. the motion of the saw is built up before it hits the bottom of the board

_____ 9. The _____ is used to make circular cuts in wood.
   A. ripsaw
   B. hacksaw
   C. compass saw
   D. miterbox

_____ 10. The _____ crosscut saw is used to cut across the grain of framing and other rough lumber.
   A. 11- or 12-point
   B. 5½-point
   C. 7- or 8-point
   D. 4½-point

## Completion

Complete each sentence by inserting the correct answer on the line near the number.

_____ 1. The term _____ is often used to denote cutting larger holes in wood.

_____ 2. The diameter of the circle made by the handle of a bit brace is known as its _____.

_____ 3. As an auger bit is turned, the _____ score the circle in advance of the cutting lips.

_____ 4. The number sixteenths of an inch in the diameter of an auger bit designate its _____.

_____ 5. The _____ is usually used to bore holes over 1" in diameter.

_____ 6. The _____ forms a recess for a flat head screw to set flush with the surface of the material in which it is driven.

_____ 7. The _____ is used to cut recesses in wood for such things as door hinges and locksets.

_____ 8. _____ come in several sizes and are used for smoothing rough surfaces and to bring work down to size.

_____ 9. The _____ is a small plane designed to be held in one hand.

_____ 10. A/An _____ is used to shape the bevel on a wood chisel or plane iron.

_____ 11. To produce a keen edge on a chisel it must be whetted using a/an _____.

_____ 12. When whetting the flat side of a chisel or plane iron, always hold the tool _____ on the stone.

_____ 13. _____ are generally used to cut straight lines on thin metal.

_____ 14. Handsaws used to cut across the grain of lumber are called _____ saws.

_____ 15. The _____ saw is used to make circular cuts in wood.

_____ 16. Coping and hacksaw blades are installed with the teeth pointing _____ from the blade.

_____ 17. The proper hacksaw blade for fast cutting in thick metal is the _____-toothed blade.

_____ 18. A saw similar to the compass saw, but designed especially for gypsum board is the _____ saw.

_____ 19. The _____ is used to cut angles of various degrees in finish lumber.

_____ 20. Ripsaws have teeth shaped like rows of tiny _____.

## Discussion

**Write your answer on the lines below each instruction.**

1. Why in a time when power tools are available, is it essential that carpenters know how to choose and skillfully use hand tools?

_____

_____

_____

2. Why is it important for the carpenter to purchase good quality tools?

_____

_____

_____

Name _____     Date _____

# CHAPTER 13   FASTENING AND DISMANTLING TOOLS

## Multiple Choice

Write the letter for the correct answer on the line next to the number of the sentence.

_____ 1. The 22-ounce framing hammer is most popularly used on _____.
   A. finish work
   B. rough work
   C. general work
   D. A, B, and C

_____ 2. The tip of a nail set is _____ to prevent it from slipping off the nail head.
   A. flattened
   B. convexed
   C. concave
   D. pointed

_____ 3. A hammer should be held _____.
   A. firmly and close to the end of the handle
   B. firmly and in the middle of the handle
   C. in different places on the handle depending on the nails you are using
   D. loosely and in the middle

_____ 4. Toenailing is a technique of driving nails _____.
   A. at an angle to fasten the end of one piece to another
   B. overhead in hard-to-reach places
   C. straight in to hardwood
   D. sideways in to end grain

_____ 5. To prevent wood from splitting or the nail from bending in hardwood, _____.
   A. presoak the wood
   B. angle the nail
   C. use hardened nails
   D. drill a hole slightly smaller than the nail shank

_____ 6. The higher the number of the point size of a Phillips screwdriver indicates _____.
   A. the higher the quality of the steel
   B. the deeper the depth of the cross shaped tip
   C. the larger the diameter of the point
   D. the lower the steel's quality

_____ 7. If a screw has a flat head _____.
   A. the shank hole must be countersunk
   B. a bit brace must be used to drive the screw
   C. a spiral screwdriver must be used
   D. a washer must be used under the head

_____ 8. The _____ is a dismantling tool available in lengths from 12"–36", that is used to withdraw spikes and for prying purposes.
   A. nail claw
   B. pry bar
   C. wrecking bar
   D. cats paw

_____ 9. C-clamp sizes are designated by _____ .
   A. their overall length
   B. the sweep of their handle
   C. their throat opening
   D. the outside length of the C

_____ 10. The proper smoothing tool to use when a considerable amount of stock is to be removed is the _____ .
   A. flat file
   B. half-round file
   C. triangular file
   D. rasp

## Completion

Complete each sentence by inserting the correct answer on the line near the number.

_____ 1. The most popular hammer for general use is the _____ .

_____ 2. As a general rule, use nails that are _____ longer than the thickness of the material being fastened.

_____ 3. To prevent the filler from falling out, it is important that finish nails are set at least _____ deep.

_____ 4. Blunting or cutting off the point of a nail helps to prevent _____ the wood.

_____ 5. In preparation for driving a screw, a shank hole and a _____ hole must be drilled.

_____ 6. When drilling the shank hole, use a _____ to prevent drilling too deep.

_____ 7. If the material to be fastened is thick, the screw may be set below the surface by _____ to gain additional penetration without resorting to a longer screw.

_____ 8. The nail claw is commonly called a _____ .

_____ 9. To turn nuts, lag screws, bolts, and other objects, an _____ is often used.

## Math

**Add the following fractions.**

1) $7/8 + 4\,3/4 =$ _____
2) $3/8 + 5/16 =$ _____
3) $11/16 + 9\,1/2 =$ _____
4) $1/8 + 5/8 =$ _____
5) $3/4 + 5\,1/16 =$ _____
6) $4\,1/2 + 9/16 =$ _____
7) $1/4 + 1/2 =$ _____
8) $7\,5/8 + 3\,1/16 =$ _____
9) $3/4 + 1/2 =$ _____
10) $2\,1/2 + 10\,3/8 =$ _____
11) $7/8 + 1/4 =$ _____
12) $3/4 + 1\,3/4 =$ _____

Name _____  Date _____

# CHAPTER 14   SAWS, DRILLS, AND DRIVERS

## Multiple Choice

**Write the letter for the correct answer on the line next to the number of the sentence.**

_____ 1. The _____ is the most used portable power tool the carpenter uses.
   A. saber saw
   B. electric circular saw
   C. heavy-duty drill
   D. reciprocating saw

_____ 2. The size of a portable circular saw is determined by _____ .
   A. its horsepower rating
   B. the length of the saw's base
   C. the saw's amperage
   D. the blade diameter

_____ 3. When operating the portable circular saw, _____ .
   A. keep the saw clear of the body until the blade has completely stopped
   B. follow the layout line closely
   C. be sure the stocks waste is over the end of the supports, not between them
   D. A, B, and C

_____ 4. Splintering occurs along the _____ with the portable circular saw.
   A. layout line of the stock
   B. opposite side of the stock from the saw
   C. beginning of the cut
   D. bottom of the kerf

_____ 5. When using the portable circular saw, the best way to prevent splintering is to _____ .
   A. lay out and cut on the finish side of the stock
   B. score the layout line with a sharp knife
   C. lay out and cut on the side opposite the finish side
   D. be sure it is on its orbital setting

_____ 6. The more teeth per inch a saber saw blade has determines _____ .
   A. the faster, but rougher it cuts
   B. the slower, but smoother it cuts
   C. the stroke range of saw
   D. if plunge cuts are posssible

_____ 7. The reciprocating saw is primarily used for _____ .
   A. cutting straight lines
   B. finish cuts
   C. bevel and miter cuts
   D. roughing-in

_____ 8. The _____ saw would be the carpenters' most likely choice for cutting holes for such things as pipes, heating and cooling ducts, and roof vents.
   A. circular
   B. saber
   C. reciprocating
   D. bayonet

_____ 9. _____ are more efficent when used in a hammer drill.
   A. Spade bits
   B. Hole saws
   C. Twist drills
   D. Masonry bits

_____ 10. The size of a portable power drill is determined by the _____.
   A. maximum opening of its chuck
   B. maximum revolutions per minute it can achieve
   C. horsepower of its motor
   D. maximum torque rating it has received

## Completion

**Complete each sentence by inserting the correct answer on the line near the number.**

_____ 1. To prevent fatal accidents from shock, make sure the tool is properly _____ and a GFCI is in the circuit being used.

_____ 2. It is important to use the proper size _____ to prevent excessive voltage drop.

_____ 3. _____ circular saw blades stay sharper longer than high speed steel blades.

_____ 4. When following a layout line with a circular saw, any deviation from the line can cause the saw to bind and possibly _____.

_____ 5. To cut out the hole for sink in a countertop, it is necessary to make a _____ cut.

_____ 6. Saber saws cut on the _____.

_____ 7. For fast cutting in wood, the reciprocating saw should be set in the _____ mode.

_____ 8. Drills that have a D-shaped handle generally are classified as _____ duty.

_____ 9. A disadvantage of a _____ is that a hole cannot be made partially through the stock.

_____ 10. Holes in metal must always be _____ to prevent the drill from wandering off center.

## Matching

Write the letter for the correct answer on the line near the number to which it corresponds.

_____ 1. GFCI  A. sometimes called a jigsaw

_____ 2. electric circular saw  B. ¼" or ⅜" chuck capacity

_____ 3. saber saw  C. used for boring holes in rough work

_____ 4. reciprocating saw  D. makes small holes in wood or metal

_____ 5. light-duty drills  E. ½" chuck capacity

_____ 6. heavy-duty drills  F. will trip a 5 milliamperes

_____ 7. twist drills  G. sometimes called a sawzall

_____ 8. spade bit  H. commonly called the skilsaw

## Discussion

Write your answer on the lines below each instruction.

1. Discuss the role that attitude plays in portable power tool safety.

   _____

   _____

   _____

2. List some general safety rules that apply to all portable power tools.

   _____

   _____

   _____

Name _____  Date _____

# CHAPTER 15  PLANES, ROUTERS, AND SANDERS

## Multiple Choice

Write the letter for the correct answer on the line next to the number of the sentence.

_____ 1. _____ can be used to sand masonry or metal.
   A. The power jointer plane
   B. The power bench plane
   C. The abrasive plane
   D. The belt sander

_____ 2. The power jointer plane can take up to _____ off the stock with one pass.
   A. ⅛"
   B. ¼"
   C. ¹⁄₁₆"
   D. ⁵⁄₁₆"

_____ 3. A major advantage of the abrasive plane is that _____ .
   A. its depth of cut is deeper than other power planes
   B. the surface is left sanded and ready for finishing
   C. it can use the same belts as the belt sander
   D. A, B, and C

_____ 4. The part of the router that holds the bit is known as the _____ .
   A. cutter head
   B. template
   C. pilot guide
   D. chuck

_____ 5. Always advance the router into the stock in a direction that is _____ .
   A. clockwise on outside edges and ends
   B. counterclockwise when making internal cuts
   C. against the rotation of the bit
   D. with the rotation of the bit

_____ 6. Extreme care must be taken while sanding when a _____ is applied.
   A. painted coating
   B. transparent coating
   C. penetrating stain
   D. A, B, and C

_____ 7. The adjusting screw on the front of a belt sander is used to _____ .
   A. retract the front roller when changing belts
   B. track the belt and center it on the roller
   C. control the belt's speed
   D. retract the belt's guard

_____ 8. When operating the belt sander, it is important to always _____.
   A. exert downward pressure on the sander
   B. sand against the grain
   C. tilt the sander in several different directions
   D. keep the electrical cord clear of the tool

_____ 9. A major disadvantage of operating the finish sander in the orbital mode is that _____.
   A. it is slower than the oscillating mode
   B. it leaves scratches across the grain
   C. it is more likely to overheat the sander
   D. the belt needs to be tracked frequently

_____ 10. Commonly used grits for finish sanding are _____.
   A. 100 or 120
   B. 60 or 80
   C. 30 or 36
   D. 12 or 16

## Completion

**Complete each sentence by inserting the correct answer on the line near the number.**

_____ 1. The power _____ is used to smoothe and straighten long edges, such as fitting doors to openings.

_____ 2. The motor on a power plane turns a _____.

_____ 3. The power _____ plane is ordinarily operated with one hand.

_____ 4. The _____ plane is similar to a power block plane, but has a heavy duty sleeve instead of a cutter head.

_____ 5. The _____ is used to make many different cuts such as grooves, dadoes, rabbets, and dovetails.

_____ 6. At least ½" of the router bit must be inserted into the router's _____.

_____ 7. Caution must always be exercised when operating the router, because the _____ is unguarded.

_____ 8. Improper use of the _____ has probably ruined more work than any other tool.

_____ 9. The most widely used sandpaper on wood is coated with _____.

_____ 10. Sandpaper _____ refers to the size of the abrasive particles.

# Math

**Subtract the following fractions.**

1) $7/8 - 3/8 =$ _____

2) $6\frac{1}{2} - 3\frac{1}{4} =$ _____

3) $13/16 - 1/4 =$ _____

4) $5\frac{1}{8} - 3/4 =$ _____

5) $1/2 - 7/16 =$ _____

6) $3/4 - 3/8 =$ _____

7) $8\frac{3}{8} - 2\frac{7}{8} =$ _____

8) $3/4 - 1/2 =$ _____

9) $10\frac{1}{2} - 7\frac{3}{4} =$ _____

10) $7/8 - 3/8 =$ _____

11) $6\frac{5}{8} - 3/4 =$ _____

12) $1/2 - 3/16 =$ _____

Name _____     Date _____

# CHAPTER 16  FASTENING TOOLS

## Multiple Choice

Write the letter for the correct answer on the line next to the number of the sentence.

_____ 1. Pneumatic fastening tools are powered by _____.
   A. explosive powder cartridges
   B. mechanical leverage
   C. compressed air
   D. disposable fuel cells

_____ 2. _____ would be used to fasten subfloor.
   A. A finish nailer
   B. A brad nailer
   C. A roofing stapler
   D. A light duty framing gun

_____ 3. Nails come _____ for easy insertion into the framing gun's magazine.
   A. glued in strips
   B. attached end to end
   C. in preloaded plastic cassettes
   D. individually packed

_____ 4. Cordless nailing guns eliminate the need for _____.
   A. air compressors
   B. long lengths of air hoses
   C. extra set-up time
   D. A, B, and C

_____ 5. A battery and spark plug is found in _____ nailing guns.
   A. pneumatic
   B. cordless
   C. powder-actuated
   D. mechanical leverage

_____ 6. The fuel cell in a cordless framing nailer can deliver enough energy to drive about _____ nails.
   A. 1200
   B. 2500
   C. 3200
   D. 4000

_____ 7. When the nailing gun's trigger is depressed, a fastener is _____.
   A. immediately driven
   B. driven when the gun's nose touches the work
   C. repeatedly ejected from the gun
   D. automatically actuated

49

_____ 8. Many states require certification to operate _____.
   A. pneumatic nailers
   B. cordless nailers
   C. powder-actuated drivers
   D. mechanical staplers

_____ 9. The strength of a powder charge can be determined by its _____.
   A. color code
   B. diameter
   C. number
   D. letter on the cartridge's base

_____ 10. In selecting a powder charge, always use _____.
   A. a charge that has more than enough power
   B. the weakest charge that will do the job
   C. a charge that will penetrate the drive pin below the surface
   D. A and C

## Completion

**Complete each sentence by inserting the correct answer on the line near the number.**

_____ 1. A nail set is not required and the possibility of marring the wood is avoided when using the _____ nailer.

_____ 2. Nails for roof nailers come in _____ of 120, which are easily loaded in a nail canister.

_____ 3. Air compressors are needed to operate _____ nailers.

_____ 4. The _____ nailer is used to fasten small moldings and trim.

_____ 5. Within the cordless nailer is an _____ engine that forces a piston down to drive the fastener.

_____ 6. A cordless finish nailer will drive about _____ nails before the battery needs to be charged.

_____ 7. A nailing gun can operate only when the work contact element is firmly _____ against the work and the trigger pulled.

_____ 8. Never leave an unattended gun with the air supply _____.

_____ 9. Specially designed pins are driven into masonry and steel by _____ drivers.

_____ 10. Never pry a powder charge out with a screwdriver or knife, this could result in an _____.

Name _____  Date _____

# CHAPTER 17 CIRCULAR SAW BLADES

## Multiple Choice

Write the letter for the correct answer on the line next to the number of the sentence.

_____ 1. The more teeth on a circular saw blade the _____ cut.
   A. faster it can
   B. rougher the surface of the
   C. smoother the surface of the
   D. longer between sharpening it can

_____ 2. If a blade is allowed to overheat, it is likely to _____ .
   A. loose its shape and wobble at high speed
   B. bind in the cut
   C. possibly cause a kickback
   D. A, B, and C

_____ 3. Coarse-tooth blades are suited for cutting _____ .
   A. thin, dry material
   B. heavy, rough lumber
   C. plastic laminated material
   D. where the quality of the cut surface is important

_____ 4. A _____ blade stays sharper longer when cutting materials that contain adhesives.
   A. carbide-tipped
   B. taper-ground
   C. planner
   D. high-speed steel

_____ 5. Every tooth on the _____ blade is filed or ground at right angles to the face of the blade.
   A. crosscut saw
   B. ripsaw
   C. combination
   D. A and C

_____ 6. The sides of all the teeth on a _____ blade are alternately filed or ground on a bevel.
   A. crosscut saw
   B. ripsaw
   C. combination
   D. B and C

_____ 7. Taper-ground blades are _____ .
   A. thinner at the tips and wider toward the center
   B. wider at the tips and thinner toward the center
   C. set to ensure the necessary clearance
   D. an excellent choice for rough, wet, heavy lumber

51

_____ 8. The ideal carbide-tipped blade for cutting plastic laminated material is the _____.
   A. triple chip grind
   B. alternative top bevel
   C. square grind
   D. combination

_____ 9. The arbor nut is always loosened by turning it _____.
   A. the opposite direction the blade rotates
   B. the same direction the blade rotates
   C. the direction of a right-handed thread
   D. the direction of a left-handed thread

_____ 10. When sharpening a circular saw blade, the process of making all the teeth the same height is known as _____.
   A. setting
   B. jointing
   C. gumming
   D. grinding

## Completion

Complete each sentence by inserting the correct answer on the line near the number.

_____ 1. Carbide-tipped blades are so hard that they must be sharpened by _____ -impregnated grinding wheels.

_____ 2. Most of the saw blades being used at present are _____.

_____ 3. The _____ saw blade is suited for when a variety of crosscutting and ripping is to be done and eliminates the need for changing blades.

_____ 4. A circular saw blade may overheat if the rate of feed is too _____.

_____ 5. The teeth on a _____ saw blade act like a series of small chisels.

_____ 6. To keep from binding, all saw blades must have some provision for _____ in the saw cut.

_____ 7. The teeth on a _____ carbide-tipped blade are similar to the rip teeth on a high speed steel blade.

_____ 8. The versatile carbide-tipped _____ blade is probably the most widely used by carpenters.

_____ 9. The valleys between the teeth on a saw blade are known as _____.

_____ 10. Arbors are always threaded in a direction that will prevent the nut from becoming _____ during operation.

## Identification: Blades

Identify each term, and write the letter of the correct answer on the line next to each number.

_____ 1. crosscut blade

_____ 2. ripsaw blade

_____ 3. combination blade

_____ 4. chisel-point

_____ 5. square grind

_____ 6. alternate bevel

_____ 7. triple chip

A.

B.

C.

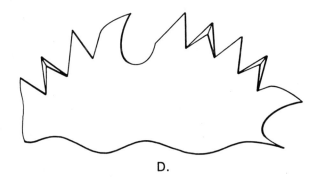
D.

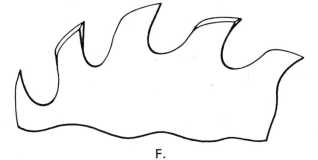
F.

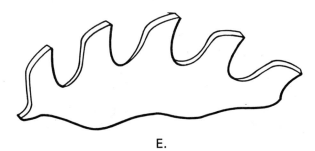
E.

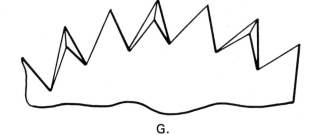
G.

53

Name _____   Date _____

# CHAPTER 18 RADIAL ARM AND MITER SAW

## Multiple Choice

Write the letter for the correct answer on the line next to the number of the sentence.

_____ 1. The size of the radial arm saw is determined by _____.
   A. the width of the board it can rip
   B. the length of the table
   C. the diameter of the largest blade it can use
   D. the horsepower of the motor

_____ 2. The depth of cut is controlled by _____.
   A. raising or lowering the table
   B. horizontally swinging the arm
   C. tilting the motor unit
   D. raising or lowering the arm

_____ 3. Adjust the depth of cut on the radial arm saw so that the saw blade is about _____ below the surface of the table.
   A. ½"
   B. ⅜"
   C. ¾"
   D. 1/16"

_____ 4. When cutting thick material on the radial arm saw, _____.
   A. pull the saw through the stock rapidly
   B. pull the saw slowly or hold it back (hesitate) somewhat
   C. never use a stop block
   D. remove the saw's guard

_____ 5. Flat miters are cut on the radial arm saw by _____.
   A. tilting the saw blade
   B. swinging the arm and simultaneously tilting the motor
   C. positioning the motor parallel to the fence
   D. rotating the arm to the desired angle

_____ 6. Compound miter cuts are frequently used on some types of _____.
   A. headers
   B. floor joists
   C. roof rafters
   D. collar ties

_____ 7. A stop block is fastened to the table of the radial arm saw when _____.
   A. ripping stock
   B. cutting many pieces the same length
   C. cutting various sized pieces
   D. when the saw is in the out-rip position

_____ 8. Upon completing a cut on the radial arm saw, always _____.
   A. raise the blade up out of the table
   B. return the saw to the starting point behind the fence
   C. place the saw directly in front of the fence
   D. lower the blade to where it contacts the table

_____ 9. The power miter saw is also called _____.
   A. the power miter box
   B. the chop saw
   C. the hew saw
   D. A and B

_____ 10. The most usual sizes of power miter saws are _____.
   A. 7½" and 8½"
   B. 6" and 8"
   C. 10" and 12"
   D. 5" and 7"

## Completion

**Complete each sentence by inserting the correct answer on the line near the number.**

_____ 1. The _____ saw is used primarily for crosscutting framing members.

_____ 2. The _____ saw is specifically designed to crosscut interior and exterior trim.

_____ 3. The arm of a radial arm saw moves horizontally in a complete _____.

_____ 4. The least obscured view when cutting miters on a radial arm saw occurs when the arm is swung to the _____.

_____ 5. The _____ locks the arm of a radial arm saw into position.

_____ 6. A/an _____ miter is cut in the same manner as a crosscut except that the radial arm saw's blade is tilted to the desired angle.

_____ 7. When ripping wide material, the motor unit is locked in the _____ position.

_____ 8. When ripping thin stock, always use a _____ to guide the stock between the blade and the fence.

_____ 9. The adjustable dado head is commonly called a/an _____.

_____ 10. In many cases, the _____ saw arrives at the job site when construction begins.

Name _____   Date _____

# CHAPTER 19   TABLE SAWS

## Multiple Choice

Write the letter for the correct answer on the line next to the number of the sentence.

_____ 1. _____ when operating the table saw.
   A. Never cut freehand
   B. Always use the rip fence when ripping
   C. Never reach over a running blade
   D. A, B, and C

_____ 2. A _____ can be cut by feeding the stock across the saw blade at an angle.
   A. dado
   B. cove
   C. rabbet
   D. V-groove

_____ 3. When the miter gauge is turned and the blade is tilted, the resulting cut is a _____.
   A. flat miter
   B. compound miter
   C. end miter
   D. bevel

_____ 4. _____ are useful aids to hold work against the fence and down on the table surface during ripping operations.
   A. Stop blocks
   B. Feather boards
   C. Taper-ripping jigs
   D. Miter jigs

_____ 5. The most common size table saw used on construction sites is the _____ model.
   A. 6″
   B. 8″
   C. 10″
   D. 12″

_____ 6. The table saw is favored over the radial arm saw for _____.
   A. ripping
   B. crosscutting
   C. miter cuts on long stock
   D. dado cuts

_____ 7. When ripping on the table saw, always use a push stick if the stock is under _____.
   A. 2″
   B. 3″
   C. 4″
   D. 5″

57

_____ 8. An auxiliary tabletop can _____ .
   A. prevent thin stock from slipping under the fence
   B. prevent narrow rippings from slipping between the saw blade and the table insert
   C. help prevent accidents
   D. A, B, and C

## Completion

**Complete each sentence by inserting the correct answer on the line near the number.**

_____ 1. The size of the table saw is determined by the diameter of the _____ .

_____ 2. The table saw's blade can be tilted up to a _____ -degree angle.

_____ 3. Handwheels on a table saw are used to adjust the _____ .

_____ 4. During the ripping operation, the stock is guided by the _____ .

_____ 5. The _____ slides in grooves on the table saw's surface.

_____ 6. Blade height is to be adjusted to about _____ above the stock.

_____ 7. A table saw operator should never stand directly in back of the _____ .

_____ 8. Cut stock that is left between a running blade and the fence may result in possible _____ that could injure those in its path.

_____ 9. Ripping on a _____ is done the same as straight ripping, except the blade is tilted.

_____ 10. When crosscutting stock, the _____ is used to guide the stock past the blade.

## Discussion

**Write your answer on the lines below each instruction.**

1. Make a list of general safety rules for the table saw.

_____

_____

_____

2. What advantages does the radial arm saw have over the table saw?

   _____

   _____

   _____

3. What advantages does the table saw have over the radial arm saw?

   _____

   _____

   _____

Name _____    Date _____

# CHAPTER 20  UNDERSTANDING DRAWINGS

## Completion

Complete each sentence by inserting the correct answer on the line near the number.

_____  1. Multi-view, also called _____ drawings, convey most of the information needed for construction.

_____  2. The lines in a/an _____ drawing diminish in size as they approach a vanishing point.

_____  3. Presentation drawings are usually the _____ type.

_____  4. The _____ plan simulates a view looking down from a considerable height.

_____  5. The direction and spacing of floor and roof framing is shown on the _____ plan.

_____  6. The drawings that show the shape and finishes of all sides of the exterior of a building are called _____ .

_____  7. Window _____ give information about the location size and type of windows to be installed in a building.

_____  8. For complex commercial projects, a _____ guide has been developed by the Construction Specifications Institute.

_____  9. The triangular _____ scale is used to scale lines when making drawings.

_____  10. The most commonly used scale on blueprints is _____ inch equals one foot.

## Matching

Write the letter for the correct answer on the line near the number to which it corresponds.

_____  1. pictorial drawings         A. take precedence over the drawing if a conflict arises

_____  2. multi-view                 B. used to show the appearance of the completed building

_____  3. isometric drawing          C. between the foot and the inch a dash is always placed

61

_____ 4. presentation drawing          D. most common ones show the kitchen and bath cabinets

_____ 5. plot plan                     E. the horizontal lines are drawn at 30° angles

_____ 6. interior elevations           F. two-dimensional drawings that convey the most information

_____ 7. modular measure               G. shows information about the lot

_____ 8. section                       H. three-dimensional isometric or perspective drawing

_____ 9. specifications                I. buildings designed using a grid with a unit of 4"

_____ 10. dimensions                   J. shows a vertical cut through all or part of a construction

## Identification: Lines

Identify each term, and write the letter of the correct answer on the line next to each number.

_____ 1. object line

_____ 2. hidden line          H. _____

_____ 3. centerline           G. ←————————→

_____ 4. section reference    F. _____

_____ 5. break line           E. - - - - - - - - - - - -

_____ 6. dimension line       D. —·—·—·—·—·—

_____ 7. extension line       C. ↑— — — — —↑

_____ 8. leader line          B. —⋀—⋀—⋀—

                                     A. ↙————

Name _____  Date _____

# CHAPTER 21  FLOOR PLANS

## Completion

Refer to Figure 21.1 to complete the following sentences.

_____  1. The overall length of the building is _____ .

_____ _____  2. The dimensions of the utility room are _____ by _____ .

_____  3. The fireplace is located in the _____ room.

_____  4. The exterior door symbol in the dining room is that of a _____ .

_____  5. The size of the floor joists in the living room and the kitchen are _____ .

_____  6. In heated areas, all exterior studs are to be _____ placed 16" on center.

_____ _____  7. The dimensions of the garage are _____ by _____ .

_____  8. Not counting the garage door, there are _____ entrance doors into the garage.

_____  9. The _____ is the only room that can be entered only through the garage.

_____  10. The plan is drawn to a scale of _____ = 1'-0".

_____  11. A _____ door is on the pantry.

_____ _____  12. The dimensions of the living room are _____ by _____ .

_____  13. In the living room, the number of electrical outlets that can be controlled by switches is _____ .

_____  14. Attic access is located in the _____ .

63

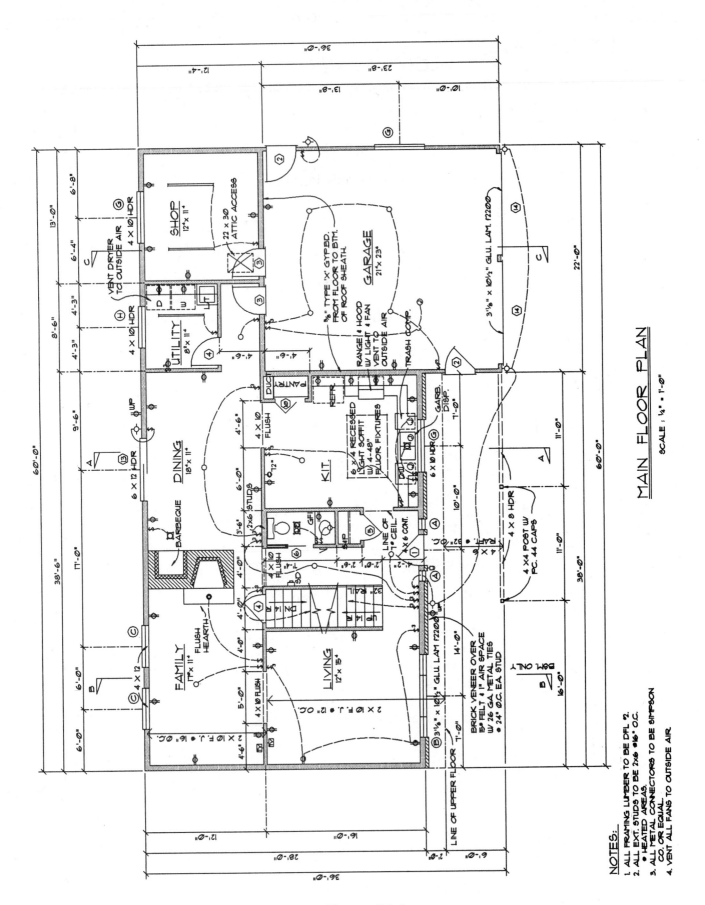

**Figure 21.1**

# Sketch: Floor Plan Symbols

**Please sketch each described item in the space provided.**

1. exterior sliding door

2. pocket door

3. bifold door

4. double-hung window

5. casement window

6. awning window

7. sliding window

8. water closet

9. standard tub

10. standard shower

## Discussion

**Write your answer on the lines below the instruction.**

1. Refer to Figure 21.1 again. If you were building the home on the floor plan for yourself to live in, what changes might you make to improve the design of the home?

_____

_____

_____

Name _____        Date _____

# CHAPTER 22   SECTIONS AND ELEVATIONS

## Completion

Base your answers to questions 1–10 on Figure 22.1.

_____   1. While floor plans are views of a horizontal cut, sections show _____ cuts.

_____   2. Sections are usually drawn at a scale of _____ = 1'-0".

_____   3. _____ are cut across the width or through the length of the entire building.

_____   4. Enlargements of part of a section, which are done to convey additional needed information, are called _____ .

_____   5. According to section A-A, the bottom of the footer must be located _____ inches below grade.

_____   6. The insulation between the crawl space and the subfloor has an R value of _____ in section A-A.

_____   7. In section A-A, a minimum distance of _____ inches must be maintained between the .006 black vapor barrier and the bottom of the floor joists.

_____   8. The insulation in the attic has a thickness of _____ inches in A-A.

_____   9. The roof has a _____-12 slope in section A-A.

_____   10. The plywood sheathing on the roof calls for a thickness of _____ in section A-A.

Base your answers to questions 11–17 on Figure 22.2.

_____   11. The rafters are framed with ____ by _____ stock in section A-B.

_____   12. The thickness of the basement floor is _____ in section A-B.

_____   13. In section A-B, the basement wall is _____ inches thick.

_____   14. Section A-B is drawn to a scale of _____" equal 1'-0".

_____   15. The insulation on the outside of the basement wall is _____ inches thick in section A-B.

_____  _____   16. In section A-B, the floor joists are framed with _____ by _____ stock.

_____   17. The footer in section A-B is _____ inch(es) thick.

67

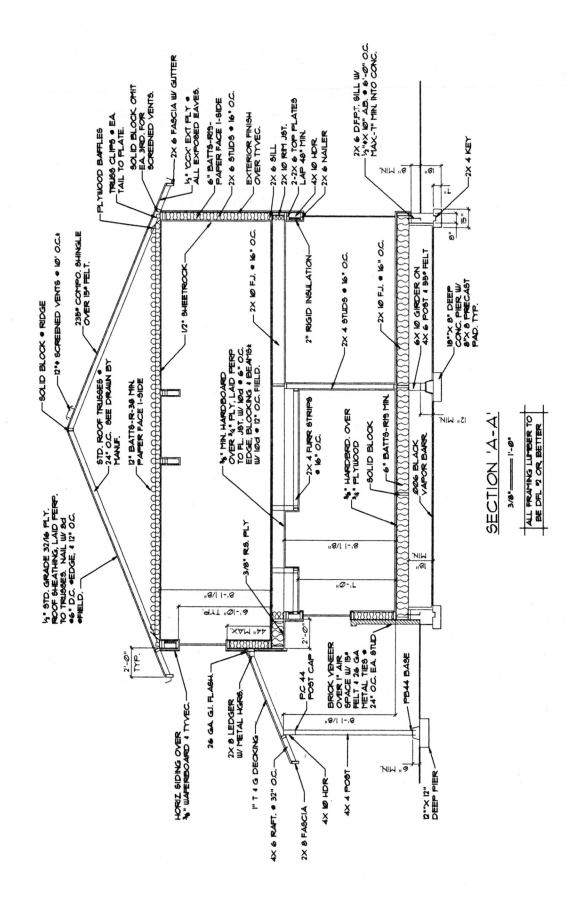

**Figure 22.1**

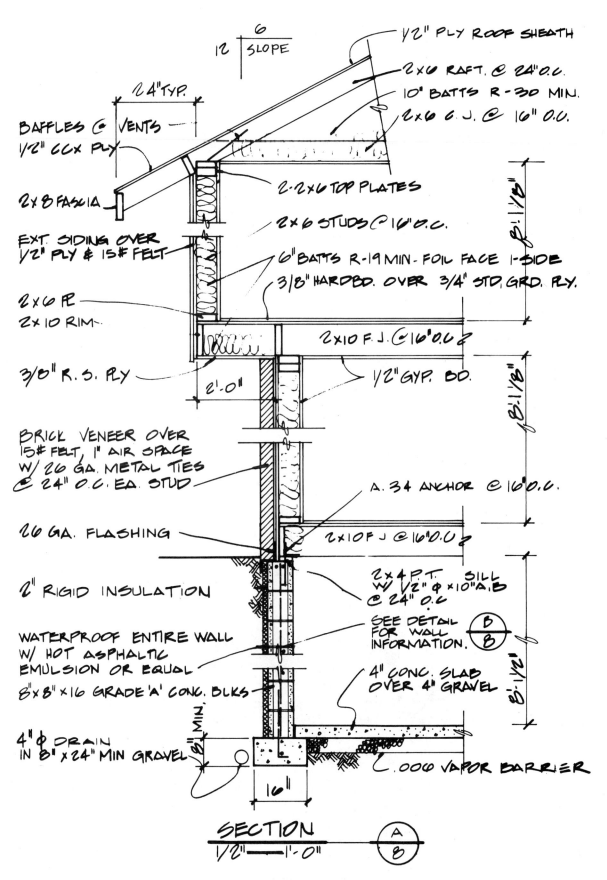

**Figure 22.2**

**Refer to Figures 22.1–22.3 to complete the following sentences.**

_____ 18. According to the accompanying detail, there are _____ inches of gravel under the concrete floor.

_____ 19. The gypsum wall board on the detail drawing has a thickness of _____ inch(es).

_____ 20. The thickness of the insulation on the detail's exterior wall is _____ inch(es).

_____ 21. Elevations are usually drawn at the same scale as the _____.

_____ 22. There are usually _____ elevations, in each set of drawings.

_____ 23. In relation to other drawings, elevations have few _____.

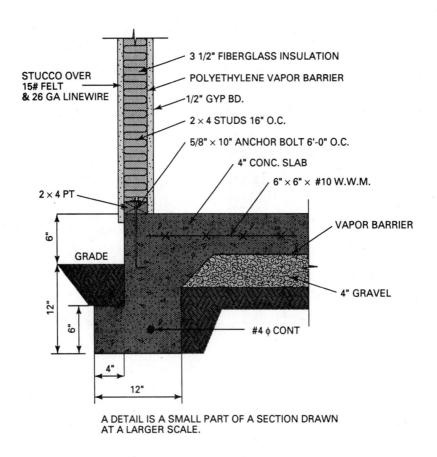

A DETAIL IS A SMALL PART OF A SECTION DRAWN AT A LARGER SCALE.

**Figure 22.3**

Name _____    Date _____

# CHAPTER 23   PLOT AND FOUNDATION PLANS

## Multiple Choice

Write the letter for the correct answer on the line next to the number of the sentence.

_____ 1. The plot plan must show _____.
   A. a frontal view of the completed structure
   B. the direction and spacing of the framing members
   C. compliance with zoning and health regulations
   D. A, B, and C

_____ 2. Metes and bounds on a plot plan refer to _____.
   A. elevation
   B. boundary lines
   C. utility easements
   D. the slope of the finish grade

_____ 3. When contour lines are spaced close together _____.
   A. a new grade is being indicated
   B. a gradual slope is indicated
   C. they are closer to sea level
   D. a steep slope is indicated

_____ 4. Instead of contour lines, the slope of the finish grade can be indicated by _____.
   A. an arrow
   B. metes and bounds
   C. easements
   D. the measurement of rods

_____ 5. The distance from the property line to the building are known as _____.
   A. the point of beginning
   B. setbacks
   C. variances
   D. bearings

_____ 6. The location of the footer on a foundation plan is indicated by _____.
   A. solid lines
   B. solid lines with a dash in the center
   C. dashed lines
   D. object lines

_____ 7. A recess in a foundation wall to support a girder is called a _____.
   A. benchmark
   B. pocket
   C. casement
   D. seat

71

_____ 8. On foundation plans, the walls are dimensioned from _____ .
   A. the centerlines of opposing walls
   B. the inside of the one wall to the inside of the next
   C. the outside area of the footer
   D. the face of one wall to the face of the next

## Completion

**Refer to Figure 23.1 to complete the following sentences.**

_____ 1. The elevation of the finished floor on the accompanying plot plan is _____ .

_____ 2. The setback distance of the home from the front boundary line is _____ on the plot plan.

_____ 3. The plot plan is drawn on a scale of 1" equals _____ feet.

_____ _____ 4. The dimensions of the home on the plot plan are _____ by _____ .

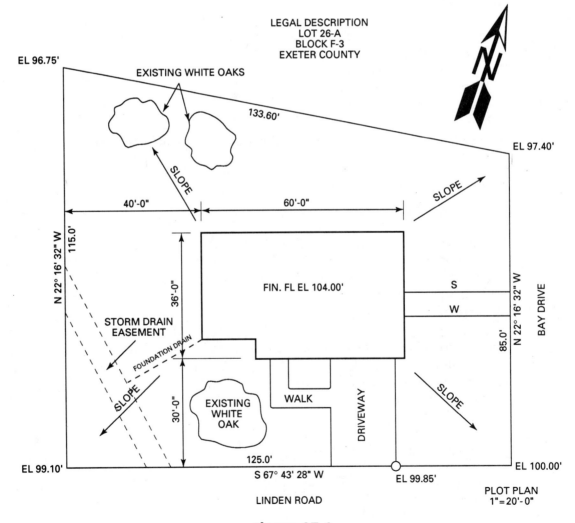

**Figure 23.1**

72

_____ 5. The longest property boundary on the plot plan is _____ feet.

_____ 6. The driveway connects to _____ Road.

_____ 7. The driveway is _____ feet long.

_____ 8. The foundation drain connects to the _____ .

_____ 9. The side of the home the water and sewer lines are connected to runs parallel to the street named _____ .

_____ 10. The front property boundary is _____ feet long.

## Math

The following elevations contain decimals of a foot. Convert the decimals of a foot to inches and sixteenths of an inch, as found on a rule.

1) 96.75'    _____          2) 99.10'    _____

3) 97.40'    _____          4) 99.85'    _____

5) 104.35'   _____          6) 93.60'    _____

7) 90.25'    _____          8) 93.10'    _____

## Discussion

**Write your answer on the lines below the instruction.**

1. What are some of the reasons municipal planning officials insist on a plot plan prior to issuance of a building permit?

_____

_____

_____

Name _____  Date _____

# CHAPTER 24  BUILDING CODES AND REGULATIONS

## Multiple Choice

Write the letter for the correct answer on the line next to the number of the sentence.

_____ 1. Similar size and purpose buildings are limited to various areas of cities and towns by _____ .
   A. building codes
   B. construction techniques
   C. zoning regulations
   D. labor unions

_____ 2. Green space refers to the _____ .
   A. minimum lot width
   B. amount of landscaped area
   C. the structure's maximum ground coverage
   D. off-street parking required

_____ 3. Structures built prior to zoning regulations that exist not within their proper zone are called _____ .
   A. unreformed
   B. nonsanctioned
   C. nonconfirming
   D. preapproved

_____ 4. Hardships imposed by zoning regulations may be relieved by a _____ granted by the zoning board.
   A. dissention
   B. objection
   C. variance
   D. reassessment

_____ 5. Minimum standards of safety concerning the design and construction of buildings are regulated by _____ .
   A. zoning laws
   B. an appeals committee
   C. construction costs
   D. building codes

_____ 6. The building code primarily used in the Northeast and Midwest is the _____ .
   A. Basic National Building Code
   B. Uniform Building Code
   C. Standard Building Code
   D. Primary Building Code

75

_____ 7. An area of importance in residential building codes is _____.
   A. exit facilities
   B. room dimensions
   C. requirements for bath, kitchens, and hot and cold water
   D. A, B, and C

_____ 8. The building permit fee is usually based on the _____.
   A. square footage of the building
   B. estimated cost of construction
   C. occupant load
   D. location of the building

_____ 9. Foundation inspections occur prior to _____.
   A. the placement of concrete
   B. the erection of forms
   C. the placement of reinforcement rod
   D. the removal of the forms

_____ 10. It is the responsibility of the _____ to notify the Building Official when it is time for a scheduled inspection.
   A. loan officer
   B. zoning officer
   C. contractor
   D. business agent

## Discussion

Write your answer on the lines below each instruction.

1. Although zoning is intended to protect the rights of the property owners, how might it at the same time infringe on these very rights?

   _____

   _____

   _____

2. Select a recent current event covered by the media where the existence of building codes has played a positive factor. Describe it below and then discuss this with your class.

   _____

   _____

   _____

3. For the most part, a good rapport exists between inspectors and builders. Why is it important for those entering the trade to be aware of this and strive to continue it?

___

___

___

Name _____ Date _____

# CHAPTER 25 LEVELING AND LAYOUT INSTRUMENTS

## Completion

Complete each sentence by inserting the correct answer on the line near the number.

_____  1. If no other tools are available, a long straightedge and a _____ may be used to level across the building area.

_____  2. An accurate tool dating back centuries, used for leveling from one point to another, is the _____ level.

_____  3. The _____ level consists of a telescope mounted in a fixed horizontal position with spirit level attached.

_____  4. Automatic levels have an internal compensator that uses _____ to maintain a true level line of sight.

_____  5. Before a level can be used, it must be placed on a _____ or some other means of support.

_____  6. When adjusting any level, never apply excessive pressure to the _____.

_____  7. The folding rule is a favored target over the tape measure because of the _____ of the tape.

_____  8. The _____ is the ideal target for longer sightings because of its clearer graduations.

_____  9. A starting point of known elevations used to determine other elevations is known as a _____.

_____  10. When the base of the rod is at the desired elevation, the reading on the rod is known as _____ rod.

_____  11. Elevation of the instrument is determined by placing the rod on the _____ and adding that reading to the elevation of the benchmark.

_____  12. When recording readings of elevation differences, all _____ sights are known as plus sights.

_____  13. When the level must be set up directly over a particular point, a _____ is attached to a hook centered below the instrument.

_____  14. A horizontal circle scale on the instrument is divided into quadrants of _____ degrees each.

79

_____ 15. The horizontal vernier is used to read _____ of a degree.

_____ 16. When laying out a horizontal angle, the instrument must be centered and leveled over the _____ of the angle.

_____ 17. By rotating a full _____ degrees, the laser level creates a level plane of light.

_____ 18. A battery-powered electronic sensor is attached to the leveling rod to detect _____ .

_____ 19. Laser level safety requires that whenever possible the laser should be set up so that it is above or below _____ level.

_____ 20. All _____ instruments are required to have warning labels attached to them.

## Matching

Write the letter for the correct answer on the line near the number to which it corresponds.

_____ 1. water level           A. minus sights

_____ 2. builder's level       B. the point of an angle

_____ 3. transit level         C. magnetic or clip on targets are attached to

_____ 4. horizontal crosshairs D. may rotate up to 40 RPS

_____ 5. vertical crosshairs   E. outside ring

_____ 6. forward sights        F. has a horizontally fixed telescope

_____ 7. horizontal circle scale G. cross-hairs used when laying out angles

_____ 8. vertex                H. limited by the length of the plastic tube

_____ 9. laser level           I. telescope can be moved up and down

_____ 10. suspended ceiling grids  J. cross-hairs used for reading elevations

Name _____  Date _____

# CHAPTER 26  LAYING OUT FOUNDATION LINES

## Multiple Choice

Write the letter for the correct answer on the line next to the number of the sentence.

_____ 1. It is usually the responsibility of the _____ to lay out building lines.
   A. mason
   B. architect
   C. carpenter
   D. foundation inspector

_____ 2. Before any layout can be made, it is important to determine _____.
   A. the dimensions of the building and its location on the site from the plot plan
   B. a starting corner by the 6-8-10 method
   C. a level plane to measure from
   D. a proper benchmark

_____ 3. Measure in on each side from the front property line the specified setback to establish the _____.
   A. approximate boundaries of the property
   B. benchmark
   C. the front line of the building
   D. different elevation points

_____ 4. In the absence of a transit or builder's level, a right triangle may be laid out using the _____.
   A. speed square
   B. water level
   C. plot plan
   D. 6-8-10 method

_____ 5. When taking measurements over layout lines, it is suggested that the tape be steadied by _____.
   A. holding it against a solid object
   B. taking readings from the 1' or 2' mark
   C. hooking the tape over the line
   D. A and B

_____ 6. If the lengths of opposite sides of a rectangular layout are equal and the diagonal measurements are also equal, then the corners are _____.
   A. parallel
   B. square
   C. divergent
   D. obtuse or acute angles

81

_____ 7. All corner stakes are located by measuring from _____.
    A. diagonal points
    B. a benchmark
    C. a point of beginning
    D. the established front and side building lines

_____ 8. _____ are wood frames to which building lines are secured.
    A. Grade rods
    B. Batter boards
    C. Line anchors
    D. Transome boards

_____ 9. Ledgers are usually _____.
    A. vertical 2" × 4" stakes
    B. horizontal 1" × 6" stakes
    C. precast concrete
    D. fastened below the top of the footer

_____ 10. Batter boards must be erected in such a manner that _____.
    A. they will last for years
    B. they will withstand sideways force when the footer is poured
    C. they will not be disturbed during excavation
    D. they can support large amounts of downward pressure.

## Measuring

**Refer to Figure 26.1 to complete the following sentences.**

_____ 1. The length of line A is _____ inches.

_____ 2. The length of line B is _____ inches.

_____ 3. The length of line C is _____ inches.

_____ 4. The length of line D is _____ inches.

_____ 5. The length of line E is _____ inches.

_____ 6. The length of line F is _____ inches.

_____ 7. The length of line G is _____ inches.

_____ 8. The length of line H is _____ inches

_____ 9. The length of line I is _____ inches.

_____ 10. The length of line J is _____ inches.

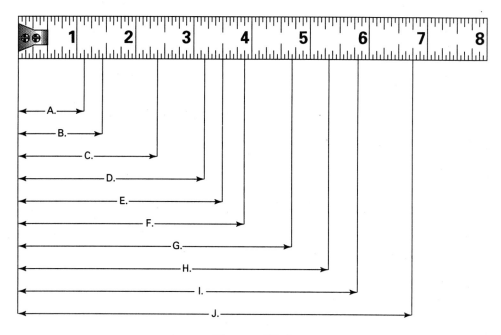

**Figure 26.1**

## Discussion

**Write your answer on the lines below each instruction.**

1. List and discuss some of the serious problems that could happen if a mistake were to occur during foundation layout.

   _____

   _____

   _____

2. How important is the knowledge of layout instruments and an understanding of plans to the foundation layout?

   _____

   _____

   _____

Name _____   Date _____

# CHAPTER 27  CHARACTERISTICS OF CONCRETE

## Multiple Choice

Write the letter for the correct answer on the line next to the number of the sentence.

_____ 1. Concrete form construction is the responsibility of the _____ .
   A. laborers
   B. masons
   C. carpenters
   D. ready-mix plant

_____ 2. A chemical reaction called _____ causes cement to harden.
   A. aggregation
   B. aeration
   C. adhesion
   D. hydration

_____ 3. Hardening of concrete can continue for _____ .
   A. days
   B. weeks
   C. months
   D. years

_____ 4. A bag of portland cement contains one cubic foot and weighs _____ pounds.
   A. 70
   B. 82
   C. 94
   D. 100

_____ 5. Type IA is an air-entraining cement used to _____ .
   A. improve resistance to freezing and thawing
   B. withstand great compression
   C. seal oil wells
   D. withstand high tempertures

_____ 6. The quality of concrete is greatly affected by _____ .
   A. the water-cement ratios
   B. rapid evaporation of the water
   C. freezing of the water
   D. A, B, and C

_____ 7. Aggregates serve as a _____ in concrete.
   A. bonding agent
   B. stabilizer
   C. filler
   D. corrosion inhibitor

85

_____ 8. A cubic yard contains _____ cubic feet.
   A. 27
   B. 81
   C. 36
   D. 18

_____ 9. Concrete must be delivered within _____ after water has been added to the mix.
   A. 30 minutes
   B. 3½ hours
   C. 15 minutes
   D. 1½ hours

_____ 10. Steel bars are added to concrete to increase its _____ .
   A. compressive strength
   B. tensile strength
   C. resistance to freezing
   D. curing time

## Completion

**Complete each sentence by inserting the correct answer on the line near the number.**

_____ 1. A #6 rebar has a diameter of _____ inch(es).

_____ 2. Welded wire mesh is identified by the gauge and spacing of the _____ .

_____ 3. To ease their removal, the inside surfaces of forms are brushed with _____ .

_____ 4. Concrete with a slump of greater than _____ inches should not be used.

_____ 5. Vibrating or hand-spading is done to eliminate voids or _____ in the concrete.

_____ 6. Excessive vibration causes concrete to become more liquid causing more _____ on forms.

_____ 7. Flooding or constant sprinkling of the surface with water is the most effective method of _____ concrete.

_____ 8. Permanent damage is almost certain if the concrete becomes _____ within the first 24 hours of being placed.

_____ 9. In cold weather, _____ are sometimes added to the concrete to shorten the setting time.

_____ 10. Concrete should be protected from freezing for at least _____ days.

## Discussion

**Write your answer on the lines below the instruction.**

1. Why is it so important for the carpenter to have a knowledge of concrete?

   _____

   _____

   _____

Name _____   Date _____

# CHAPTER 28  FORMS FOR SLABS, WALKS, AND DRIVEWAYS

## Multiple Choice

Write the letter for the correct answer on the line next to the number of the sentence.

_____ 1. Slab-on grade construction permits the structure to have _____ .
   A. lower construction costs
   B. a basement
   C. a crawl space
   D. a higher profile

_____ 2. For good drainage with slab-on grade construction the top of the slab should be _____ .
   A. not more than 4" below grade
   B. level with the grade
   C. not more than 4" above grade
   D. not less than 8" above grade

_____ 3. The soil under a slab is sometimes treated with chemicals to _____ .
   A. reduce settling
   B. prevent frost from lifting the slab
   C. control termites
   D. improve drainage

_____ 4. Prior to pouring the slab for a slab-on grade structure, _____ .
   A. a vapor barrier should be installed
   B. all water and sewer lines must be installed
   C. top soil must be removed
   D. A, B, and C

_____ 5. With a monolithic slab _____ .
   A. no footer is necessary
   B. the slab and footer are one piece
   C. a basement is included
   D. no reinforcement is needed in the slab

_____ 6. In an area where the ground freezes to an appreciable depth, the type of slab-on grade construction that must be used is known as _____ .
   A. a monolithic slab
   B. an independent slab
   C. a detached slab
   D. a thickened edge slab

_____ 7. To provide for expansion and contraction of the slab, ridged insulation is placed _____.
   A. under the slab
   B. between the slab edge and the foundation wall
   C. on the perimeter of the foundation wall
   D. on top of the slab

_____ 8. Forms for walks and driveways should be built so that _____.
   A. water will drain from their surfaces
   B. they may be left in place to save on labor cost
   C. the concrete will rest on sod when possible
   D. A, B, and C

_____ 9. When building forms for curved walks or driveways, it is helpful to _____.
   A. use hardwood
   B. install the grain horizontally if using plywood
   C. wet the stock before bending
   D. prestress the form for several days prior to the pour

_____ 10. When bending curves with a long radius _____ may be used for forms.
   A. 2" × 4"s
   B. ridged insulation
   C. 1" × 4"s
   D. A and B

## Completion

Complete each sentence by inserting the correct answer on the line near the number.

_____ 1. The _____ for a foundation provides a base to spread the load of the structure over a wide area of the soil.

_____ 2. To provide support for posts and columns a _____ footing is used.

_____ 3. In residential construction, the width of the footing is usually _____ the thickness of the wall.

_____ 4. Reinforcement rods of specified size and spacing are placed in footings of larger buildings to increase their _____.

_____ 5. The depth of footings must be below the _____ line.

_____ 6. If the soil is stable, form work is not necessary and the concrete can be carefully placed in a _____ of proper width and depth.

_____ 7. When constructing forms, the use of _____ nails ensures easy removal.

_____ 8. _____ are nailed to the top edges of the form to tie them together and keep them from expanding.

_____  9. A _____ is formed in a footing by pressing 2" × 4" lumber into freshly poured concrete.

_____  10. Sometimes it is necessary to _____ the footing if it is built on sloped land.

_____  11. Another name for a thickened edge slab is a _____ slab.

_____  12. Forms for walks and driveways should be built so that water will _____ from the concretes' surface.

_____  13. When forming curves with ¼" plywood, install it with the grain _____ for easier bending.

_____  14. _____ the stock also sometimes helps the bending process when forming curves.

## Discussion

**Write your answer on the lines below each instruction.**

1. There is an old saying among builders, "If you don't have a good foundation you don't have anything." Explain the reasoning behind this statement.

   _____

   _____

   _____

2. Why is it so important to place a footer below the frost level?

   _____

   _____

   _____

Name _____   Date _____

# CHAPTER 29  WALL AND COLUMN FORMS

## Multiple Choice

Write the letter for the correct answer on the line next to the number of the sentence.

_____ 1. To increase the efficiency of forming foundation walls, _____ .
   A. the forms are custom-built in place
   B. panels and panel systems are used
   C. snap ties are eliminated
   D. ⅝" plywood is used to construct the walers

_____ 2. A usual _____ size is 36" wide by 8' high.
   A. waler
   B. strongback
   C. girder pocket
   D. panel

_____ 3. Snap ties are used to _____ .
   A. hold the wall forms together at the desired distance
   B. support the wall forms against the lateral pressure of the concrete
   C. reduce the need for external bracing
   D. A, B, and C

_____ 4. The projecting ends of snap ties are snapped off _____ .
   A. prior to placing the concrete
   B. inside the concrete after the removal of the forms
   C. slightly protruding from the concrete after it is placed
   D. when erecting the panels

_____ 5. Walers are _____ .
   A. spaced at right angles to the panel frame members
   B. always ran horizontally
   C. constructed of ⅝" plywood
   D. A, B, and C

_____ 6. The higher a concrete wall, _____ .
   A. the more lateral pressure on the top of the form
   B. the less lateral pressure at the bottom of the form
   C. the fewer snap ties needed at the bottom of the form
   D. the greater the lateral pressure on the bottom of the form

_____ 7. To provide a smooth face to the hardened concrete and for easy stripping of the forms, _____ .
   A. all panel faces should be oiled or treated with a chemical-releasing agent
   B. the panels should be placed horizontally
   C. the concrete must not be vibrated
   D. it is recommended that the panels not be placed on plates

_____ 8. When concrete walls are to be reinforced, the rebars are _____.
   A. installed after the walers are spaced
   B. added after the concrete is placed
   C. used instead of snap ties
   D. tied in place before the inside panels are erected

_____ 9. A _____ is a thickened portion of the wall added for strength or support for beams.
   A. strongback
   B. pilaster
   C. gusset
   D. parapet

_____ 10. Anchor bolts are set in the wall _____.
   A. after the concrete has partially set
   B. at the same time as the rebar
   C. as soon as the wall is screeded
   D. when the concrete is placed

_____ 11. It is important that anchor bolts be _____.
   A. staggered and placed at various heights
   B. attached to the rebar
   C. set at the correct height and at the specified locations
   D. A and B

_____ 12. A _____ is a form that provides an opening in a foundation wall for things such as ducts, pipes, doors, and windows.
   A. domeform
   B. sleeper
   C. keyway
   D. buck

## Completion

Complete each sentence by inserting the correct answer on the line near the number.

_____ 1. A radius can be formed on the corners of a concrete column by fastening _____ molding to the panel edges.

_____ 2. Quarter-round molding can be attached to the panels to form a _____ shape.

_____ 3. By attaching triangular-shaped strips of wood to the edges of a panel, a _____ is formed on the column's corners.

_____ 4. A column may be decorated with flutes by attaching vertical strips of _____ molding spaced on the panel faces.

_____ 5. _____ are often used to provide the face of the column with various textures, such as wood, brick, and stone.

_____ 6. The number and spacing of yokes depends on the _____ of the column.

_____ 7. With column form construction, vertical _____ are installed between the overlapping ends of the yokes.

_____ 8. A concrete _____ consists of manufactured items for concrete form construction.

# CHAPTER 30  STAIR FORMS

## Completion

Complete each sentence by inserting the correct answer on the line near the number.

_____ 1. To conserve on concrete when forming stairs, it may be necessary to lay out the stairs before the _____ is placed.

_____ 2. When forming earth-supported stairs between two existing walls, the _____ and the _____ are laid out on the inside of the existing walls.

_____ 3. Planks are ripped to width to correspond to the height of each _____, when forming stairs.

_____ 4. When forming stairs, it is important to _____ the bottoms of the planks used to form the risers. This permits the mason to trowel the entire edge of the tread.

_____ 5. Riser planks are braced from top to bottom between their ends to keep them from _____ due to the concrete's pressure.

_____ 6. Suspended stairs must be designed and reinforced to support not only their own weight but also the weight of the _____ .

_____ 7. On suspended stairs with open ends, the layout is made on the _____ used for forming the stairs' ends.

_____ 8. Short lengths of narrow boards known as _____ are fastened across joints in forms to strengthen them.

_____ 9. A _____ is a thin piece of plywood that has the width of the tread and the height of the riser laid out on it. It is used to mark the tread and riser locations on the form.

_____ 10. Economical concrete construction depends a great deal on the _____ of form.

## Measuring

Refer to Figure 30.1 to complete the following sentences.

_____ 1. The length of line A is _____ inches.

_____ 2. The length of line B is _____ inches.

_____ 3. The length of line C is _____ inches.

_____ 4. The length of line D is _____ inches.

_____ 5. The length of line E is _____ inches.

_____ 6. The length of line F is _____ inches.

_____ 7. The length of line G is _____ inches.

_____ 8. The length of line H is _____ inches

_____ 9. The length of line I is _____ inches.

_____ 10. The length of line J is _____ inches.

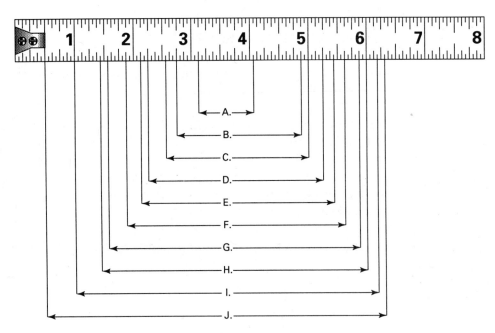

Figure 30.1

Name _____  Date _____

# CHAPTER 31  TYPES OF FRAME CONSTRUCTION

## Multiple Choice

Write the letter for the correct answer on the line next to the number of the sentence.

_____ 1. The most widely used framing method in residential construction is _____.
   A. balloon frame construction
   B. platform frame construction
   C. post and beam frame construction
   D. the Arkansas system

_____ 2. A platform frame is easy to construct because _____.
   A. the second floor joists rest on 1" × 4" ribbon
   B. at each level a flat surface is provided to work on
   C. the studs run uninterruptedly the entire height of the building
   D. it uses fewer but larger pieces

_____ 3. Lumber shrinks mostly _____.
   A. across its width
   B. across its thickness
   C. from end to end
   D. A and B

_____ 4. A relatively large amount of settling occurs in _____.
   A. platform frame construction
   B. balloon frame construction
   C. post and beam frame construction
   D. B and C

_____ 5. In balloon frame construction, it is important that _____.
   A. firestops be installed in the walls at several locations
   B. the second floor platform is erected on top of the walls of the first floor
   C. brick or stucco not be used to finish the outside walls
   D. interior design be planned around the supporting roof beam posts

_____ 6. Wood frame construction is used for residential and light commercial construction because _____.
   A. the cost is usually less than other types
   B. it provides for better insulation
   C. it is very durable and will last indefinitely if properly maintained
   D. A, B, and C

_____ 7. A post and beam roof is usually _____.
   A. not insulated
   B. insulated on top of the deck
   C. insulated below the deck
   D. more labor- and material-consuming than conventional roof framing

97

_____ 8. House depths that are not evenly divided by four _____.
   A. conserve material
   B. waste material
   C. save on labor costs
   D. A, and C

_____ 9. Reducing the clear span of floor joists _____.
   A. makes it necessary to use larger joists
   B. can be accomplished by using narrower sill plates
   C. can make it possible to use smaller-sized joists
   D. can always result in higher costs if wider sill plates must be used

_____ 10. An energy-saving construction system that uses 2" × 6' wall studs spaced 24" on center is called _____.
   A. the Arkansas system
   B. the 24" Module Method
   C. the western frame
   D. post and beam construction

## Discussion

**Write your answer on the lines below each instruction.**

1. Explain in detail the reason it is important that building dimensions be divisible by four.

   _____

   _____

   _____

2. Given a choice between platform, balloon, or post and beam construction in a home you would be building for yourself, which would you choose and why?

   _____

   _____

   _____

Name _____ Date _____

# CHAPTER 32   LAYOUT AND CONSTRUCTION OF THE FLOOR FRAME

## Multiple Choice

Write the letter for the correct answer on the line next to the number of the sentence.

_____ 1. Heavy beams that support the inner ends of the floor joists are called _____.
   A. sill plates
   B. girders
   C. subflooring
   D. bridging

_____ 2. If the blueprints do not specify the kind or size of structural component _____.
   A. the job foreman may have to estimate this data
   B. always oversize the material to be safe
   C. have the design checked by a professional engineer
   D. consult with an insurance expert

_____ 3. Pockets should be large enough to provide at least a _____ bearing for the girder.
   A. 12"
   B. 1½"
   C. 8"
   D. 4"

_____ 4. Anchor bolts are used to attach the _____ to the foundation.
   A. floor joists
   B. sill plate
   C. girder
   D. A, B, and C

_____ 5. When tightening the nuts on anchor bolts, be sure to _____.
   A. replace the washers first
   B. not overtighten them
   C. tighten them as tight as possible
   D. A and B

_____ 6. In conventional framing, floor joists are usually placed _____.
   A. 16" on center
   B. prior to the girder
   C. 18" on center
   D. below the sill plate

_____ 7. Notches in the bottom or top of sawn lumber floor joists should not _____.
   A. exceed ⅜ the joist depth
   B. be in the middle third of the joist span
   C. be in the first third of the joist span
   D. exceed ½ the joist depth

99

_____ 8. When measuring for floor joist layout, it is best to _____.
   A. mark the joist locations on top of the foundation
   B. measure and mark each joist individually
   C. have the ends of the plywood fall directly on the edge of the joists
   D. use a tape stretched along the length of the building

_____ 9. Joists are installed with _____.
   A. the crown down
   B. the lap spiked together over the sill plate
   C. the crown up
   D. A and B

_____ 10. When installing plywood subflooring _____.
   A. leave a 1/16" space at all panel end joints
   B. leave an 1/8" between panel edges
   C. all end joints are made over joists
   D. A, B, and C

## Completion

**Complete each sentence by inserting the correct answer on the line near the number.**

_____ 1. Sill plates lie directly on the foundation wall and provide bearing for the _____.

_____ 2. If joists are lapped over the girder, the minimum amount of lap is _____ inches.

_____ 3. Holes bored in joists for piping or wiring should not be larger than _____ of the joist depth.

_____ 4. Shortened floor joists at the ends of floor openings are called _____ joists.

_____ 5. It is important that the floor joist layout permits the plywood sheets to fall directly on the _____ of the joists.

_____ 6. The area of a rectangular floor is determined by multiplying the length by its _____.

_____ 7. To determine the number of rated panels of subflooring required, divide the floor area by _____.

_____ 8. A one-story home with the dimensions of 28' × 48' would require _____ rated panels for the subfloor.

_____ 9. To determine the total linear feet of wood cross-bridging needed, multiply the length of the building by _____ for every row of bridging.

## Identification: Floor Frame

Identify each term, and write the letter of the correct answer on the line next to each number.

_____ 1. girder

_____ 2. floor joist

_____ 3. rim joists

_____ 4. sill plate

_____ 5. foundation wall

_____ 6. column footing

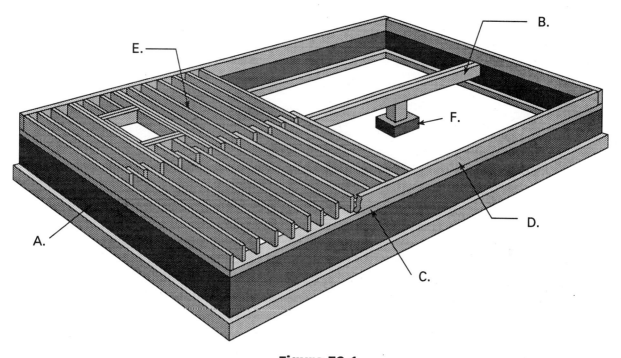

**Figure 32.1**

## Math

The following questions concern estimating the necessary material to construct a floor frame on a foundation measuring 28' × 40'.

1. Using 2" × 6"s for the sill plate, how many board feet must be ordered?

**101**

2. If the floor joists are placed 16" on center and none are to be doubled, how many are needed?

3. How many sheets of plywood are needed for the subfloor?

Name _____  Date _____

# CHAPTER 33 CONSTRUCTION TO PREVENT TERMITES AND FUNGI

## Multiple Choice

Write the letter for the correct answer on the line next to the number of the sentence.

_____ 1. For effective treatment against dry wood termites, it is usually necessary to _____.
   A. install earth-to-wood termite barriers, and chemically treat the foundation and soil
   B. tent-fumigate the whole home with a toxic gas
   C. eliminate any moisture that is reaching the wood and allow the wood to dry out
   D. spray the infested area with over-the-counter insecticides

_____ 2. The most destructive species of termite is _____.
   A. the dry wood termite
   B. the damp wood termite
   C. the subterranean termite
   D. the northern pine termite

_____ 3. Protection against subterranean termites should be considered _____.
   A. at the first sign of an infestation
   B. when planning and during construction of a building
   C. prior to occupation of the building
   D. after the elimination of the queen

_____ 4. Prevention of termite attacks is based on _____.
   A. keeping the wood dry
   B. making it as difficult as possible for termites to get to the wood
   C. chemical treatment of the soil
   D. A, B, and C

_____ 5. In crawl spaces, clearance between the ground and the bottom of the floor joists should be at least _____.
   A. 8"
   B. 12"
   C. 14"
   D. 18"

_____ 6. The _____ foundation provides the home the best protection against termites.
   A. slab-on grade
   B. monolithic slab
   C. two core concrete block
   D. independent slab

103

_____ 7. Wall siding should not extend more than _____ below the top of the foundation wall.
   A. 2"
   B. 4"
   C. 6"
   D. 8"

_____ 8. Research has shown that due to _____ termite shields have not been effective in preventing infestations.
   A. improper installation
   B. the failure to frequently inspect for signs of shelter tubes
   C. the shields' rapid deterioration
   D. A and B

_____ 9. To safely work with pressure treated lumber, it is important _____.
   A. to wear eye protection and a dust mask when sawing or machining it
   B. upon completion of the work to wash your hands before eating and drinking
   C. not to burn the leftover scraps
   D. A, B, and C

_____ 10. Clothing that accumulates sawdust from pressure treated wood should be _____.
   A. disposed of
   B. treated with a disinfectant
   C. laundered separately from other clothing and before reuse
   D. washed in boiling water then dried outdoors

## Completion

Complete each sentence by inserting the correct answer on the line near the number.

_____ 1. _____ termites tend to cause less damage to buildings than other types of termites.

_____ 2. Discovery of a leak sometimes reveals a _____ termite infestation.

_____ 3. All wood scraps should be _____ from the area before backfilling around the foundation.

_____ 4. A pit in the ground filled with stones used to absorb drainage water is called a _____.

_____ 5. Cracks as little as _____ of an inch permit the passage of termites.

_____ 6. Wall siding should be at least _____ above the finish grade.

_____ 7. _____ grade pressure treated lumber is used for sill plates, joists, girders and decks.

_____ 8. Termites will not usually eat treated wood, however they will _____ over it to reach wood that is not treated.

Name _____  Date _____

# CHAPTER 34 EXTERIOR WALL FRAME PARTS

## Multiple Choice

Write the letter for the correct answer on the line next to the number of the sentence.

_____ 1. It is important for the carpentry student to be able to know the _____ of the different parts of the wall frame.
   A. names
   B. functions
   C. locations
   D. A, B, and C

_____ 2. Names given to different parts of a structure _____.
   A. are the same nationwide
   B. may differ with the geographical area
   C. always include the size of the framing member
   D. change if engineered lumber is used

_____ 3. The bottom horizontal member of a wall frame is called the _____.
   A. sole plate
   B. top plate
   C. double plate
   D. sill plate

_____ 4. Vertical members of the wall frame that run full length between the plates are known as _____.
   A. ribbons
   B. trimmers
   C. braces
   D. studs

_____ 5. It is necessary that headers _____.
   A. be installed wherever interior partitions meet exterior walls
   B. be left in until the wall section is erected
   C. be strong enough to support the load over the opening
   D. be placed under the rough sill

_____ 6. When headers are built to fit a 4" wall, they consist _____.
   A. of two pieces of 2" lumber
   B. of three pieces of 2" lumber
   C. of two pieces of 2" lumber with ½" plywood or strand board sandwiched in between
   D. of two pieces of 2" lumber with ¾" plywood or strand board sandwiched in between

105

_____ 7. Many carpenters prefer to use double 2" thickness rough sills _____ .
   A. to bear the weight over the window
   B. to cut down on insulation
   C. to provide more surface to nail window trim
   D. to fasten the drywall to

_____ 8. Shortened studs that carry the weight of the header are called _____ .
   A. trimmers
   B. door jacks
   C. braces
   D. A and B

_____ 9. Ribbons are horizontal members that support the second floor _____ .
   A. partition intersections
   B. joists in balloon construction
   C. corner posts
   D. headers

_____ 10. Generally, corner bracing is not needed if _____ is used on the corners.
   A. softboard
   B. Celotex
   C. insulated board sheathing
   D. rated panel wall sheathing

## Completion

Complete each sentence by inserting the correct answer on the line near the number.

_____ 1. Studs are usually _____ , unless 6" insulation is desired in the exterior wall.

_____ 2. Studs are usually spaced _____ or 24" on center.

_____ 3. The depth of the header depends on the _____ of the opening.

_____ 4. Headers in _____ walls are made up of three pieces of 2" lumber and two pieces of ½" plywood or strand board.

_____ 5. The use of _____ lumber for headers permits the spanning of wide openings that otherwise might need additional support.

_____ 6. Rough sills are members used to form the bottom of a _____ opening.

_____ 7. Most of the time, corner posts are constructed of _____ 2" × 4"s.

_____ 8. Ribbons are usually made of _____ stock.

_____ 9. When needed, corners are braced using 1" × 4"s that are set into the face of the studs, top plate, and sole plate, and ran _____ .

# Identification: Exterior Wall Frame Parts

Identify each term, and write the letter of the correct answer on the line next to each number.

_____ 1. corner post

_____ 2. top plate

_____ 3. sole plate

_____ 4. stud

_____ 5. jack stud

_____ 6. corner brace

_____ 7. partition intersection

_____ 8. trimmer

_____ 9. rough sill

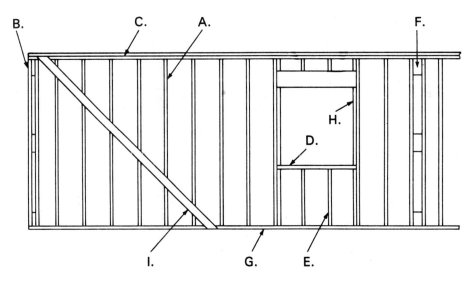

**Figure 34.1**

107

Name _____  Date _____

# CHAPTER 35  FRAMING THE EXTERIOR WALL

## Multiple Choice

Write the letter for the correct answer on the line next to the number of the sentence.

_____ 1. To determine stud length, the carpenter must know the _____.
A. thickness of the finished floor
B. thickness of the ceiling below the ceiling joists
C. height from the finish floor to the ceiling
D. A, B, and C

_____ 2. The carpenter must determine _____ sizes from information contained in the door and window schedules.
A. header
B. corner post
C. rough opening
D. A and B

_____ 3. When laying out a rough opening, a clearance of ____ is usually maintained between the door frame and the rough opening.
A. ¼"
B. ½"
C. ¾"
D. 1"

_____ 4. The sides and tops of a door frame are called _____.
A. stops
B. rabbets
C. jambs
D. door sets

_____ 5. Interior partitions on blueprints are usually _____.
A. dimensioned to their centerlines
B. shown on the plot plan
C. eliminated
D. represented by dotted lines on the floor plan

_____ 6. When laying out studs, _____.
A. vary their spacing between 8" and 10"
B. keep the studs directly in line with the joists below
C. mark their location on the sill plate
D. A, B, and C

**109**

_____ 7. Prior to bracing the end sections of a wall, an easy and effective way to check the wall section for square is to _____ .
   A. use a framing square
   B. measure the corner posts to see if they are the same
   C. measure the wall from corner to corner both ways to see if they are the same
   D. check the window openings with a combination square

_____ 8. For accurate plumbing of corner posts use _____ .
   A. a transit level
   B. a plumb bob
   C. a 6' level with accessory aluminum blocks attached to each end
   D. A, B, and C

_____ 9. A job-built combination wall aliner and brace is called a _____ .
   A. spring brace
   B. wall jack
   C. aliner strut
   D. partition brace

_____ 10. All softboard sheathing is fastened to the wall using _____ nails.
   A. 6d
   B. 12d
   C. duplex
   D. roofing

## Completion

**Complete each sentence by inserting the correct answer on the line near the number.**

_____ 1. A _____ is an opening framed in the wall in which to install doors and windows.

_____ 2. The bottom member of a door frame is called a _____ .

_____ 3. Consulting the manufacturers' _____ is the best way to determine window rough openings.

_____ 4. The first step in laying out wall openings is to consult the blueprints to find the _____ dimension of all the openings.

_____ 5. To avoid problems when installing finish work, it is important that all edges of frame members are kept _____ wherever they join each other.

_____ 6. All studs in a wall should have their crowned edges facing _____ when the wall is erected.

_____ 7. When applying temporary bracing to an erected wall, use _____ nails or drive the nails only partway into the lumber.

_____ 8. Partitions that carry a load are referred to as _____ partitions.

_____ 9. Wall _____ covers the exterior walls.

_____ 10. When estimating material for an exterior wall with a layout that is 16" on center, figure one _____ for each linear foot.

## Math

**The following problems concern estimating the necessary material needed to frame the outside walls for a building that measures 24' by 32'. The studs are placed 16" on center and are 2" × 4"s.**

1. How many linear feet of 2" × 4"s should be ordered for the plates?

2. How many studs should be ordered?

3. If the wall is 8' high, how many sheets of fiber board sheathing will be needed to cover it?

Name _____   Date _____

# CHAPTER 36  CEILING JOISTS AND PARTITIONS

## Multiple Choice

Write the letter for the correct answer on the line next to the number of the sentence.

_____ 1. Because of the added weight on bearing walls, it is required that _____.
   A. the top plate be doubled
   B. the sole plate be doubled
   C. they be erected after the roof is on
   D. they be erected in a manner different than exterior walls

_____ 2. Roof trusses eliminate the need for bearing walls because _____.
   A. they distribute all the weight on the interior walls
   B. the girder carries the weight
   C. they transmit the weight to the exterior walls
   D. they weigh less than conventional roof framing

_____ 3. Rough openings width for interior doors is equal to _____.
   A. the door width plus 1″ and the thickness of the finish floor
   B. the door width plus 1″ and twice the thickness of the door stop
   C. the door width plus 1″ and twice the thickness of the door jamb
   D. the width of the door plus 1¾″

_____ 4. At exterior walls, ceiling joists are placed so that _____.
   A. the sole plate is attached to them
   B. the rafters can be attached to their sides
   C. they will support the roof trusses until they are set into place
   D. A and C

_____ 5. When ceiling joists are installed in line, _____.
   A. their ends must butt together at the center line of the bearing partition
   B. a scab must be installed at the splice
   C. the exterior wall location for either side of the span is on the same side of the rafter
   D. A, B, and C

_____ 6. The ends of ceiling joists on the exterior wall must be cut _____.
   A. if additional head room is needed
   B. slightly above the rafter at the same angle as the roof's slope
   C. flush or slightly below the top edge of the rafter
   D. only if the roof has a very steep pitch

_____ 7. When low-pitched hip roofs are used, _____.
   A. roof trusses are a necessity
   B. ceiling joists must be doubled
   C. stub joists must be installed
   D. ceiling joists may be eliminated

113

_____ 8. Ceiling joists on each end of the building are placed so that _____.
   A. the outside face is flush with the inside of the wall
   B. the outside face is flush with the outside of the wall
   C. there is a 1½″ gap between the joists and outside wall
   D. there is access for the electrician to run wiring through them

_____ 9. When metal lath or ceiling tiles are to be used, it is necessary to _____.
   A. alter the ceiling joist spacing to accommodate the size of the ceiling material
   B. use powder-actuated drivers to attach them to the joist
   C. attach the material directly to the ceiling joist
   D. first install furring strips

_____ 10. Most building codes state that the studs of nonbearing partitions may be spaced up to _____ on center.
   A. 16″
   B. 24″
   C. 28″
   D. 32″

## Completion

Complete each sentence by inserting the correct answer on the line near the number.

_____ 1. Ceiling joists tie the exterior side walls together and provide a base for the ceiling _____.

_____ 2. Size and spacing of ceiling joists are determined by the _____ or from local building codes.

_____ 3. _____ prevent the roof from exerting outward pressure on the wall, which would cause the walls to spread.

_____ 4. When joists are installed, be sure the crowned edges are pointed _____.

_____ 5. When cutting the taper on the ends of ceiling joists, be sure the taper does not exceed _____ times the depth of the member.

_____ 6. A _____ is a small opening in ceiling joists that allows access to the attic.

_____ 7. Joists are to be toenailed into the plates with at least two _____ penny nails.

_____ 8. The installation of _____ enables the electrician to run wires at right angles to the joists without boring holes in each one.

_____ 9. Because nonbearing partitions carry no load, headers are usually doubled _____.

_____ 10. Bathroom and kitchen walls sometimes must be made thicker to accommodate _____ later installed in those walls.

_____ 11. A rough opening width of _____″ is needed for a 32″ interior door, if the jamb stock is ½″ thick.

Name _____    Date _____

# CHAPTER 37  BACKING, BLOCKING, AND BASES

## Completion

Complete each sentence by inserting the correct answer on the line near the number.

_____  1. A short block of lumber that is installed in floor, wall, and ceiling cavities to provide fastening for various fixtures and parts is known as _____ .

_____  2. The placement of blocking and backing is not usually found in a set of _____ .

_____  3. Wall blocking is needed at the _____ edges of wall sheathing panels permanently exposed to the weather.

_____  4. Blocking is required between studs in walls over _____ high.

_____  5. It is easier to install blocking in a _____ line than in a straight one.

_____  6. The white mineral mined from the earth that plaster is made from is called _____ .

_____  7. The thin wood strips that in the past were used for a plaster base are known as _____ .

_____  8. _____ and metal lath are the plaster bases that are commonly used today.

_____  9. Gypsum lath that is 3/8" thick may be used if the framing is spaced ___" on center.

_____  10. The standard width of gypsum lath is 16" and the standard length is ___".

_____  11. Metal lath is formed from sheet metal that has been slit and _____ to form numerous small openings to key the plaster.

_____  12. Specialists called _____ install metal lath on large commercial jobs.

_____  13. Gypsum lath requires _____ coats of plaster.

_____  14. To prevent cracks, metal lath called _____ is installed at inside corners for reinforcement.

_____  15. All exterior corners must have _____ of expanded metal installed on them.

_____  16. Plasterers use _____ as guides to control the plaster's thickness.

115

## Identification: Blocking and Backing

Identify each term, and write the letter of the correct answer on the line next to each number.

_____ 1. faucet backing

_____ 2. tub support blocking

_____ 3. showerhead backing

_____ 4. outlet backing

_____ 5. shower curtain rod backing

_____ 6. lavatory backing

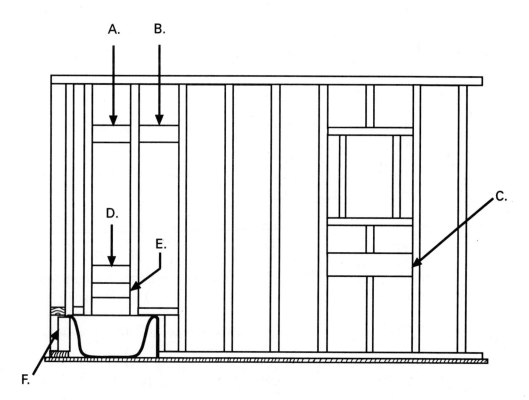

**Figure 37.1**

Name _____   Date _____

# CHAPTER 38  METAL FRAMING

## Completion

Complete each sentence by inserting the correct answer on the line near the number.

_____  1. All steel framing members are coated with material that resists _____ .

_____  2. Studs for interior nonload-bearing applications are manufactured from 25-, 22-, and _____ -gauge steel.

_____  3. Pipes and conduit can be ran through punchouts that are located at intervals in the studs _____ .

_____  4. The top and bottom horizontal members of a steel frame wall are called _____ .

_____  5. Steel channels are used in suspended ceilings and for _____ of walls.

_____  6. One and a quarter inch oval head screws are used to fasten runners into _____ .

_____  7. To cut metal framing to length, tin snips can be used on _____ gauge steel.

_____  8. Maximum spacing for metal furring channels is _____ inches on center.

## Discussion

Write your answer on the lines below each instruction.

1. What are some of the factors that could lead to steel framing members being used more frequently in the future?

   _____

   _____

   _____

2. What usually determines whether specialists in light steel or carpenters do the framing?

   _____

   _____

   _____

3. Over an extended period of time, which type of stud—steel or wood—is more environmentally favorable? Explain your answer.

_____

_____

_____

Name _____    Date _____

# CHAPTER 39  WOOD, METAL, AND PUMP JACK SCAFFOLDS

## Multiple Choice

Write the letter for the correct answer on the line next to the number of the sentence.

_____ 1. All scaffolds must be capable of supporting without failure at least _____.
    A. the maximum intended load
    B. two times the maximum intended load
    C. three times the maximum intended load
    D. four times the maximum intended load

_____ 2. Single-pole wooden scaffolds are used when _____.
    A. block basements are being built
    B. they can be attached to the wall with no interference to the work
    C. a continuous comfortable working height is desired
    D. chimneys are being erected above the roof line

_____ 3. Vertical members of a scaffold are called _____.
    A. ledgers
    B. bearers
    C. braces
    D. poles

_____ 4. To prevent excessive checking, scaffold planks _____.
    A. should be painted
    B. should not exceed 6′ in length
    C. should have their ends banded with steel
    D. be replaced yearly

_____ 5. All scaffold planks must be _____.
    A. scaffold grade or its equivalent
    B. be laid with their edges close together
    C. should not overhang the bearer by more than 12″
    D. A, B, and C

_____ 6. On scaffolds that are more than ten feet high, _____ are installed on all open sides.
    A. bearers
    B. outriggers
    C. bucks
    D. guardrails

_____ 7. Double-pole scaffolds should be set _____.
    A. up tight against the wall
    B. as far from the wall as the worker can comfortably reach
    C. as close to the wall as is possible without interfering with the work
    D. without braces on the inside face on medium duty ones

_____ 8. If the height of a double-pole scaffold exceeds 25', _____.
   A. its capacity should be limited to two workers
   B. all scaffold planks must be doubled
   C. workers should wear safety harnesses
   D. it must be secured to the wall at intervals of 25'

_____ 9. With metal scaffolding _____.
   A. only one set of braces is needed
   B. braces must be forced on to fit correctly
   C. each section consists of two end pieces and two folding braces
   D. more time is needed to erect it, due to its difficulty to work with

_____ 10. Pump jack scaffolds should not be used when _____.
   A. the working level exceeds 500 lbs
   B. more than two people are needed to do the job
   C. the poles exceed 30' in height
   D. A, B, and C

## Completion

**Complete each sentence by inserting the correct answer on the line near the number.**

_____ 1. Wood scaffolds are designated as light, medium, or heavy duty according to the _____ they are required to support.

_____ 2. In single-pole staging, bearers not attached to the corners of the building must be fastened to a notched _____.

_____ 3. _____ are diagonal members that stiffen the scaffolding and prevent the poles from moving or buckling.

_____ 4. On wooden scaffolds, the top guardrail is usually 2" × 4" lumber and fastened to the poles about _____ inches above the working platform.

_____ 5. Between poles, ledgers are never to be _____.

_____ 6. Scaffold planks should be of equal lengths so that the ends are _____ with each other.

_____ 7. If the length of a double-pole scaffold exceeds 25', it must be secured to the _____ at intervals not greater than 25' horizontally.

_____ 8. When erecting metal scaffolding, always level it until the _____ fit easily.

_____ 9. Casters for mobile scaffolding must support _____ times the maximum intended load.

_____ 10. On pump jack scaffold poles, the braces must be installed at vertical intervals not exceeding _____ feet.

## Identification: Scaffold Parts

Identify each term, and write the letter of the correct answer on the line next to each number.

_____ 1. scaffold pole

_____ 2. bearer

_____ 3. top rail

_____ 4. mid-rail

_____ 5. toe board

_____ 6. ledger

_____ 7. scaffold plank

_____ 8. wall ledger

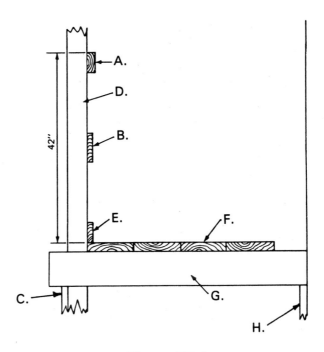

**Figure 39.1**

Name _____   Date _____

# CHAPTER 40   BRACKETS, HORSES, AND LADDERS

## Completion

**Complete each sentence by inserting the correct answer on the line near the number.**

_____   1. Wall brackets are bolted to a short block of 2″ × 4″ lumber placed against at least two _____ on the inside wall.

_____   2. If wall bracket staging is over 10′ in height, _____ must be installed.

_____   3. Roof brackets are usually required whenever the roof has more than a ____ inch vertical rise per horizontal foot of run.

_____   4. Roof brackets should be placed at about _____ foot horizontal intervals.

_____   5. When nailing roof brackets, use three 8d common nails driven home; try to get at least one nail into a _____ .

_____   6. A _____ is a low working platform supported by a bearer with spreading legs at each end.

_____   7. For light duty work, horses and trestle jacks should not be spaced more than _____ feet apart.

_____   8. If a horse scaffold is arranged in tiers, no more than _____ tiers should be used.

_____   9. The base of an extension ladder should be held a distance out from the wall equal to _____ the ladder's vertical height.

_____   10. When used to reach a roof or a working platform, the top of the ladder must extend at least _____ feet above the top support.

_____   11. Always be careful of overhead electric lines when using a _____ ladder.

_____   12. Metal brackets installed on ladders to hold scaffold planks are called _____ .

_____   13. A typical saw horse is 36″ wide with 24″ _____ .

_____   14. Work stools used by finish carpenters to support work or serve as step stools are also called _____ .

_____   15. _____ are another name for ladder rungs.

123

## Discussion

**Write your answer on the lines below each instruction.**

1. What are some of the things those responsible for erecting scaffolds must be aware of?

   _____

   _____

   _____

2. Do you think it would be advisable for construction firms to use professional scaffold erectors? If so, when and why?

   _____

   _____

   _____

Name _____   Date _____

# CHAPTER 41  ROOF TYPES AND TERMS

## Matching

Write the letter for the correct answer on the line near the number to which it corresponds.

_____ 1. gable roof                 A. the wider and longer side

_____ 2. shed roof                  B. uppermost horizontal line of the roof

_____ 3. hip roof                   C. usually the width of the building

_____ 4. gambrel roof               D. inches of rise per foot of run

_____ 5. span                       E. two sloping roofs meeting at the top

_____ 6. rafter                     F. slopes in one direction only

_____ 7. total run                  G. sloping member of the roof frame

_____ 8. ridge                      H. vertical when the rafter is in position

_____ 9. total rise                 I. slopes upward from all walls of the building to the top

_____ 10. line length               J. vertical distance that the rafter rises

_____ 11. pitch                     K. horizontal when the rafter is in position

_____ 12. slope of the roof         L. gives no consideration to the thickness of the stock

_____ 13. plumb cut                 M. horizontal travel of the rafter

_____ 14. level cut                 N. a variation of the gable roof

_____ 15. framing square            O. the amount of slope of the roof

## Identification: Roofs

Referring to the figure on the following page, identify each term and write the letter of the correct answer on the line next to each number.

_____ 1. gable roof            _____ 4. intersecting roof

_____ 2. shed roof             _____ 5. gambrel roof

_____ 3. hip roof              _____ 6. mansard roof

125

A.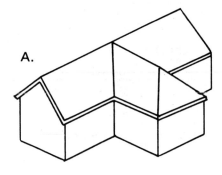

B.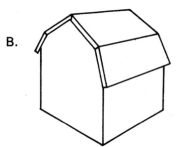

C.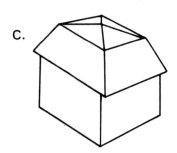

D.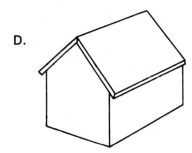

E.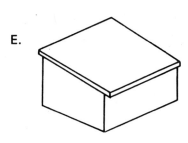

F.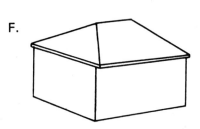

## Discussion

**Write your answer on the lines below each instruction.**

1. Describe the line length method of calculating rough rafter lengths. Include a sketch of the process in your description.

   _____

   _____

   _____

2. When would the above process be acceptable to use? When would it not be acceptable to use?

   _____

   _____

   _____

Name _____  Date _____

# CHAPTER 42 GABLE AND GAMBREL ROOFS

## Completion

Complete each sentence by inserting the correct answer on the line near the number.

_____ 1. The most common style of roof used is the _____ roof.

_____ 2. On an equal-pitched gable roof, the _____ rafter is the only type of rafter needed to be laid out.

_____ 3. Although not absolutely necessary, the _____ simplifies roof erection.

_____ 4. When laying out the rafter that is to be used as a pattern, be sure to select the _____ piece possible.

_____ 5. When using the step-off method to determine the length of a common rafter, the framing square must be moved _____ times if the width of the building is 28′.

_____ 6. Rafter tables come in booklet form and are also stamped on one side of a _____.

_____ 7. On roofs with moderate slopes, the length of the level cut of the seat is usually the width of the _____.

_____ 8. In most plans, the rafter overhang is given in terms of a _____ measurement.

_____ 9. If a ridge board is used, the rafter must be _____ a distance equal to ½ the width of the ridge board.

_____ 10. All ridge board joints must be centered on a _____.

_____ 11. End rafters are commonly called _____ rafters.

_____ 12. When an overhang is required at the rakes, horizontal structural members called _____ must be installed.

_____ 13. When installing _____ , care must be taken not to force the end rafters up and create a crown in them.

_____ 14. Usually gambrel roof rafters meet at a continuous member called a _____ .

## Identification: Gable and Gambrel Roof Parts

Identify each term, and write the letter of the correct answer on the line next to each number.

_____ 1. ridge

_____ 2. common rafter

_____ 3. plumb cut or ridge cut

_____ 4. seat cut or bird's mouth

_____ 5. tail or overhang

_____ 6. collar ties

_____ 7. gable studs

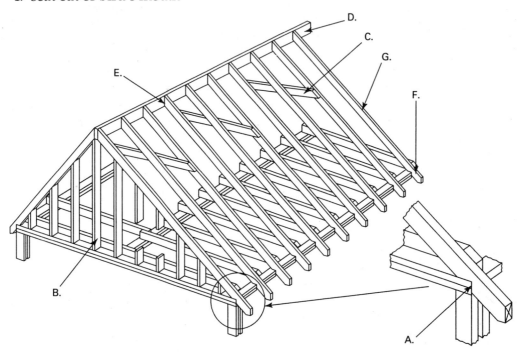

## Math

The following rafter line lengths contain decimals of a foot. Convert the decimals of a foot to inches and sixteenths of an inch as found on a rule.

1. 12.37'   12' 4 7/16
2. 19.22'   19' 2 10/16
3. 17.69'   17' 8 4/16
4. 14.76'   14' 9 2/16
5. 13.42'   13' 5
6. 8.95'    8' 11 6/16
7. 13.40'   13' 4 12/16
8. 21.63'   21' 7 9/16

128

Name _____   Date _____

# CHAPTER 43   HIP ROOFS

## Multiple Choice

**Write the letter for the correct answer on the line next to the number of the sentence.**

_____ 1. Hip roofs are _____ .
   A. easier to frame than gable roofs
   B. more complicated than gable roofs to frame
   C. the most common style of roof used
   D. sloped in one direction only

_____ 2. Hip rafters are required _____ .
   A. where the slopes of a hip roof meet
   B. on all roof framing
   C. on saltbox style roofs
   D. at right angles from the plates to the common rafters

_____ 3. In comparison to the common rafters on the same roof, _____ .
   A. hip rafters must rise the same but with fewer steps
   B. hip rafters have a decreased unit of run
   C. the slope of a hip rafter is much steeper
   D. the unit of run of the hip rafter is increased

_____ 4. If the pitch of a hip roof is a 6" rise per foot of run, the hip rafter would be laid out by holding the square at _____ .
   A. 6 and 12
   B. 6 and 14
   C. 6 and 17
   D. 6 and 24

_____ 5. The ridge cut of a hip rafter _____ .
   A. is a compound angle
   B. is called the cheek cut
   C. may be called a side cut
   D. A, B, and C

_____ 6. When finding the length of a hip rafter using the tables on the framing square, _____ .
   A. the figure found square is divided by the overall run of the rafter
   B. the figures from the second line are used
   C. the common difference must first be determined
   D. A, B, and C

_____ 7. When laying out the seat cut of a hip rafter, _____ .
   A. consideration must be given to fitting it around the corner of the wall
   B. consideration must be given to dropping the hip rafter
   C. the next to the last plumb line laid out is used as the plumb cut for the seat of the rafter
   D. the figures from the third line of the framing square are used

_____ 8. When the corners of a hip rafter are beveled flush with the roof, it is called _____.
   A. raising the hip
   B. dropping the hip
   C. backing the hip
   D. opposing the hip

_____ 9. A double-cheek cut is usually made at _____.
   A. the tail cut of a hip rafter
   B. the seat cut of a hip rafter
   C. the seat cut of a hip-jack rafter
   D. the tail cut of a hip-jack rafter

_____ 10. The hip-jack rafter _____.
   A. runs at right angles from the hip rafter to the plate
   B. has the same unit of run as the hip rafter
   C. is actually a shortened common rafter
   D. has a double-cheek cut

_____ 11. The hip-jack rafter meets the hip roof at _____.
   A. 22½°
   B. 45°
   C. 60°
   D. 90°

_____ 12. The line length of a hip roof ridge is _____.
   A. the same as the overall length of the building
   B. found by subtracting ½ the width of the building from its length
   C. the length of the building minus its width
   D. determined from information on the rafter tables of the framing square

## Identification: Hip Roof Parts

Identify each term, and write the letter of the correct answer on the line next to each number.

_____ 1. hip jack rafter        _____ 4. hip rafter

_____ 2. ridge                  _____ 5. plate

_____ 3. common rafter

## Discussion

**Write your answer on the lines below the instruction.**

1. When planning a home design, what factors might go into the decision of choosing a hip roof?

   _____

   _____

   _____

Name _____  Date _____

# CHAPTER 44  INTERSECTING ROOFS

## Multiple Choice

Write the letter for the correct answer on the line next to the number of the sentence.

_____ 1. Intersecting roofs must always have _____ .
   A. hip rafters
   B. valley rafters
   C. purlins
   D. kneewalls

_____ 2. If the heights of two intersecting roofs are different, _____ .
   A. supporting and shortened valley rafters must be used
   B. a valley cripple jack rafter must be used
   C. valley-jack rafters must be used
   D. A, B, and C

_____ 3. The unit of run for valley rafters is _____ .
   A. 12
   B. the same as common rafters
   C. called the minor span
   D. 17

_____ 4. In order for the valley rafter to clear the inside corner of the wall, _____ .
   A. it must be dropped like the hip rafter
   B. the seat cut must be extended
   C. the seat cut must be raised
   D. the wall plate is notched

_____ 5. The supporting valley rafter is shortened _____ .
   A. by ½ the thickness of the ridge board
   B. by the thickness of the ridge board
   C. by ½ the 45° thickness of the ridge board
   D. only if a shortened valley rafter is needed

_____ 6. The total run of the supporting valley rafter is _____ .
   A. the run of the common rafter of the main roof
   B. the run of the common rafter of the smaller roof
   C. known as the major span
   D. A and C

_____ 7. The tail cut of the valley rafter is _____ .
   A. identical to that of a hip rafter
   B. is a single cheek cut
   C. a square cut
   D. a double cheek cut that angles inward

133

_____ 8. The unit of run used to determine the length of the valley-jack rafter is _____.
   A. the same one used for valley rafters
   B. the same one used for hip rafters
   C. the same one used for common rafters
   D. 17" if the step-off method is used

_____ 9. The total run of any valley-jack rafter is _____.
   A. the same as the common rafters
   B. known as the minor span
   C. that of the common rafter minus the horizontal distance it is located from the corner of the building
   D. that of the supporting valley rafter minus the horizontal distance it is located from the end of the building

_____ 10. Hip valley cripple-jack rafters cut between the same hip and valley rafters _____.
   A. differ in size by the same common difference found on the rafter table
   B. have square cheek cuts
   C. are the same length
   D. A and B

## Identification: Intersecting Roof Parts

Identify each term, and write the letter of the correct answer on the line next to each number.

_____ 1. ridge of major span

_____ 2. supporting valley rafter

_____ 3. shortened valley rafter

_____ 4. common rafter

_____ 5. ridge of minor span

_____ 6. valley cripple-jack rafter

_____ 7. hip-jack rafter

_____ 8. valley-jack rafter

_____ 9. hip valley cripple-jack rafter

_____ 10. hip rafter

## Discussion

**Write your answer on the lines below the instruction.**

1. What are the three main things to be remembered that will help eliminate confusion concerning the layout of so many different types of rafters?

   _____

   _____

   _____

Name _____   Date _____

# CHAPTER 45 SHED ROOFS, DORMERS, AND SPECIAL FRAMING PROBLEMS

## Completion

Complete each sentence by inserting the correct answer on the line near the number.

_____ 1. The shed roof slopes in only _____ direction.

_____ 2. The unit of run for a shed roof is _____ inches.

_____ 3. A _____ is a framed projection above the plane of the roof containing one or more windows.

_____ 4. If not using the step-off method, the length of a shed roof rafter can be determined by using the _____ rafter tables.

_____ 5. Shed roofs are framed by _____ the rafters into the plate at the designated spacing.

_____ 6. When framing a shed roof, it is important that the plumb cut of the seat be kept snug against the _____.

_____ 7. The rafters on both sides of a dormer opening must be _____.

_____ 8. Top and bottom _____ of sufficient strength must be installed when dormers are framed with their front wall partway up the main roof.

_____ 9. In most cases, shed dormer roofs extend to the ridge of the main roof in order to gain enough _____.

## Discussion

Write your answer on the lines below the instruction.

1. What is the main disadvantage of building an intersecting roof after the main roof has been framed and sheathed?

_____

_____

_____

137

Name _____     Date _____

# CHAPTER 46  TRUSSED ROOFS

## Multiple Choice

Write the letter for the correct answer on the line next to the number of the sentence.

_____ 1. Truss use in roof framing _____ .
  A. increases the home's usable attic space
  B. slows down the home's construction
  C. eliminates the need for load-bearing interior partitions below
  D. A, B, and C

_____ 2. The diagonal parts of roof trusses are called _____ .
  A. web members
  B. knee supports
  C. stringers
  D. sleepers

_____ 3. The upper cords of a roof truss act as _____ .
  A. ceiling joists
  B. rafters
  C. collar ties
  D. a ridge board

_____ 4. Most trusses used today _____ .
  A. are designed by the carpenter
  B. are made in fabricating plants
  C. are built on the job
  D. do not contain gusset plates

_____ 5. Approved designs and instructions for job-built trusses are available from the _____ .
  A. American Plywood Association
  B. Truss Plate Institute
  C. American Hardwood Association
  D. A and B

_____ 6. The most common truss design is the _____ truss.
  A. howe
  B. pratt
  C. fink or W
  D. scissors

_____ 7. Carpenters are more involved in the _____ of trusses.
  A. erection
  B. design
  C. construction
  D. fabrication

139

_____ 8. Failure to properly erect and brace a trussed roof could result in _____.
   A. collapse of the structure
   B. loss of life or serious injury
   C. loss of time and material
   D. A, B, and C

_____ 9. The truss bracing system depends a great deal on _____.
   A. the use of 1" × 4" temporary bracing
   B. scabs nailed to the end of the building
   C. how well the first truss is braced
   D. 8d duplex finish nails

_____ 10. As bracing is installed, it is important that _____.
   A. exact spacing be maintained
   B. spacing be continually readjusted
   C. it is only applied to two planes of the truss assembly
   D. it has a maximum length of no more than 8'

## Completion

Complete each sentence by inserting the correct answer on the line near the number.

_____ 1. The plane of the top cord is known as the _____ plane.

_____ 2. The plane of the bottom cord is known as the _____ plane.

_____ 3. It is recommended that continuous lateral bracing be placed within _____ inches of the ridge on the top cord plane.

_____ 4. Diagonal bracing between the rows of lateral bracing should be placed on the _____ of the top plane.

_____ 5. Continuous lateral bracing of the bottom cord must be applied to maintain the proper _____.

_____ 6. Bottom cord bracing is nailed to the _____ of the bottom cord.

_____ _____ 7. Bottom cord bracing should be installed at intervals no greater than _____ to _____ feet along the width of the building.

_____ 8. Rated panels of plywood and _____ are commonly used to sheath roofs.

_____ 9. In post and beam construction where roof supports are spaced further apart, _____ is used for roof sheathing.

_____ 10. Panel clips, tongue and groove edges, or other adequate blocking must be used when _____ exceed the indicated value of the plywood roof sheathing.

140

## Discussion

**Write your answer on the lines below each instruction.**

1. List the advantages of using trusses over conventional roof framing.

   _____

   _____

   _____

2. What is the major disadvantage to using roof trusses? How might this affect future remodeling or sale of the home?

   _____

   _____

   _____

Name _____     Date _____

# CHAPTER 47  STAIRWAYS AND STAIRWELLS

## Multiple Choice

Write the letter for the correct answer on the line next to the number of the sentence.

_____ 1. The tread run is _____ .
   A. the part of the tread that extends beyond the riser
   B. the horizontal distance between the faces of the risers
   C. the finish material that covers the vertical distance from one step to another
   D. a nonskid material applied to tread

_____ 2. The housed finish stringer is _____ .
   A. usually fabricated in a shop
   B. usually built on the job site
   C. always used on service stairs
   D. installed by the framing crew

_____ 3. Dadoes routed into the sides of the housed finished stringer _____ .
   A. reduce the stair's rise
   B. increase the headroom
   C. support and house the risers and treads
   D. A, B, and C

_____ 4. The preferred angle for ease of stair climbing is between _____ degrees.
   A. 20 and 25
   B. 25 and 30
   C. 30 and 35
   D. 35 and 40

_____ 5. To determine the individual rise, the _____ must be known.
   A. total run
   B. total rise
   C. riser thickness
   D. tread thickness

_____ 6. To determine riser height without mathematics, _____ used.
   A. a sliding T bevel and a level are
   B. the common rafter table on the framing square is
   C. a story pole and a set of dividers are
   D. a plumb bob and a line level are

_____ 7. The sum of one rise and one tread should equal _____ .
   A. between 17 and 18
   B. between 18 and 19
   C. between 19 and 20
   D. less than 17

_____ 8. Decreasing riser height _____.
   A. decreases the run of the stairs
   B. increases the run of the stairs
   C. uses less space
   D. makes the stairs more difficult to climb

_____ 9. Increasing riser height _____.
   A. decreases the run of the stairs
   B. makes the stairs more difficult to climb
   C. uses less space
   D. A, B, and C

_____ 10. Most building codes call for a minimum of _____ of headroom for stairways.
   A. 6'-2"
   B. 6'-4"
   C. 6'-6"
   D. 6'-8"

## Completion

**Complete each sentence by inserting the correct answer on the line near the number.**

_____ 1. The stairwell is framed at the same time as the _____.

_____ 2. _____ stairs extend from one habitable level of the house to another.

_____ 3. _____ stairs extend from a habitable level to a nonhabitable level.

_____ 4. In residential construction, stairways should not be less than _____ inch(es) wide.

_____ 5. A _____ stairway is continuous from one floor to another without any turns or landings.

_____ 6. Stairs that have intermediate landings between floors are called _____ stairs.

_____ 7. An L-type platform stair changes direction _____ degrees.

_____ 8. A platform stairway that changes directions 180° is called a _____ -type stairway.

_____ 9. A _____ staircase gradually changes directions as it ascends from one floor to another.

_____ 10. A _____ stairway is constructed between walls.

_____ 11. _____ stairways have one or both sides open to a room.

_____ 12. The vertical distance between finish floors is the total _____ of the stairway.

_____ 13. The total horizontal distance that the stairway covers is known as the total _____ .

_____ 14. The vertical distance from one step to another is the _____ .

_____ 15. The opening in the floor that the stairway passes through is the _____ .

## Math

**Determine the riser height for straight stairways that have the following total rises.**

1. 8'-7"      _____

2. 8'-5 ½"    _____

3. 8'-10 ¾"   _____

145

Name _____ Date _____

# CHAPTER 48  STAIR LAYOUT AND CONSTRUCTION

## Multiple Choice

Write the letter for the correct answer on the line next to the number of the sentence.

_____ 1. When laying out stair carriages, make sure _____.
   A. riser heights are greater than tread widths
   B. all riser heights are the same
   C. all tread widths are the same
   D. B and C

_____ 2. When scaling across a framing square to determine the rough length of a stair carriage, use the side of the square that is graduated in _____ of an inch.
   A. sixteenths
   B. eighths
   C. tenths
   D. twelfths

_____ 3. When stepping off the stair carriage with a framing square, _____.
   A. stand on the side of the crowned edge
   B. lay out the rise with the square's tongue
   C. lay out the tread run with the square's blade
   D. A, B, and C

_____ 4. To be sure the bottom riser is the same height as all the other risers it may be necessary to _____.
   A. add blocking under the stair carriage bottom
   B. cut a certain amount off the stair carriage bottom
   C. alter the thickness of the first tread
   D. alter the thickness of all treads but the first

_____ 5. The top riser is equalized by _____.
   A. lowering and cutting the level line at the top of the stair carriage
   B. altering the top treads thickness
   C. fastening the stair carriage at the proper height in relation to tread and finish floor thickness
   D. adding blocking to the top of the stair carriage

_____ 6. Residential staircases of average width usually _____.
   A. require three carriages
   B. only need two carriages
   C. are less than 36" wide
   D. are more than 48" wide

147

_____ 7. Built-up stair carriages are used to _____.
   A. simplify drywall application
   B. conserve wood
   C. avoid dropping the stair carriage
   D. increase the stairs' strength

_____ 8. If drywall is applied after the stairs are framed, _____.
   A. repair or remodeling work becomes more difficult than if it were applied before framing
   B. time is saved
   C. blocking is required between studs in back of the stair carriage
   D. no blocking is needed

_____ 9. U-type stairways usually have the landing _____ of the stairs.
   A. near the bottom
   B. in the middle
   C. near the top
   D. A and C

_____ 10. According to most codes, any flight of stairs that has a vertical distance of 12' or more must have _____.
   A. double railings on each side
   B. a width of at least 4'
   C. at least one landing
   D. a tread run of at least 12"

## Discussion

**Write your answer on the lines below each instruction.**

1. What important safety rules must be considered when laying out stairs?

   _____

   _____

   _____

2. Although winder stairs are not recommended, what are some of the rules that must be followed if they are used?

   _____

   _____

   _____

Name _____   Date _____

# CHAPTER 49  THERMAL AND ACOUSTICAL INSULATION

## Multiple Choice

Write the letter for the correct answer on the line next to the number of the sentence.

_____ 1. Organic insulation materials are treated to make them resistant to _____ .
   A. fire
   B. insects
   C. vermin
   D. A, B, and C

_____ 2. _____ insulation would be a good choice for insulating the sidewalls of an older, uninsulated home.
   A. Loose-fill
   B. Rigid
   C. Flexible
   D. Reflective

_____ 3. Rigid insulation is usually made of _____ .
   A. vermiculite
   B. glass wool
   C. fiber or foamed plastic material
   D. rock wool

_____ 4. Reflective insulation should be installed _____ .
   A. tight against the siding
   B. facing an air space with a depth of ¾" or more
   C. in contact with the foundation wall when used in basements or crawl spaces
   D. only in homes without air conditioning

_____ 5. Foamed-in-place insulation _____ .
   A. expands on contact with the surface
   B. is made from organic fibers
   C. contains glass wool
   D. is commonly used for sheathing and decorative purposes

_____ 6. To reduce heat loss, all _____ that separate heated from unheated areas must be insulated.
   A. walls
   B. ceilings
   C. roofs and floors
   D. A, B, and C

_____ 7. When installing flexible insulation between floor joists over crawl spaces, _____.
   A. the vapor barrier faces the heated area
   B. the vapor barrier faces the ground
   C. remove the vapor barrier
   D. at least a ¾" air space must be maintained between the joists and the insulation

_____ 8. The resistance to the passage of sound through a building section is rated by its _____.
   A. Sound Transmission Class
   B. Impact Noise Rating
   C. Sound Absorption Class
   D. Impact Transmission Class

## Completion

Complete each sentence by inserting the correct answer on the line near the number.

_____ 1. _____ insulation prevents the loss of heat in cold seasons and resists the passage of heat into air-conditioned areas in the hot seasons.

_____ 2. _____ insulation reduces the passage of sound from one area to another.

_____ 3. Proper ventilation must be provided within the building to remove _____ that forms in the space between the cold surface and the thermal insulation.

_____ 4. If confined to a small space in which it is still, _____ is an excellent insulator.

_____ 5. Aluminum foil is used as an insulating material that works by _____ heat.

_____ 6. The higher the R-value number of insulation, the more _____ the material is.

_____ 7. The use of _____ studs in exterior walls permits the installation of 6" insulation.

_____ 8. _____ insulation is manufactured in blanket and batt form.

_____ 9. The _____ on both sides of blanket insulation facing are used for fastening it to studs or joists.

_____ 10. Batt insulation is available in thicknesses of up to _____ inches.

Name _____  Date _____

# CHAPTER 50  CONDENSATION AND VENTILATION

## Multiple Choice

**Write the letter for the correct answer on the line next to the number of the sentence.**

_____ 1. When the temperature of moisture-laden air drops below its dew point, _____ occurs.
   A. evaporation
   B. condensation
   C. vaporization
   D. dehydration

_____ 2. Condensation of water vapor in walls, attics, roofs, and floors _____.
   A. increases the R-value of any insulation it comes in contact with
   B. prevents the wood from rotting
   C. leads to serious problems
   D. only occurs in warmer climates

_____ 3. To prevent the condensation of moisture in a building, _____.
   A. reduce the moisture in the warm, inside air
   B. ventilate the attic
   C. install a barrier to the passage of water vapor to areas that cannot be ventilated
   D. A, B, and C

_____ 4. In a well-ventilated area, _____.
   A. condensed moisture is removed by evaporation
   B. moisture is forced back into the insulation
   C. insulation is not necessary
   D. condensation is at its worst

_____ 5. Ventilation of air space on the cold side of insulation is especially difficult _____.
   A. in the attic
   B. in a fully insulated wall
   C. in crawl spaces
   D. A, B, and C

_____ 6. On roofs where the ceiling finish is attached to the rafters and insulation is installed, _____.
   A. an air space is not needed
   B. a moisture barrier should be placed on the cold side of the insulation
   C. a well-vented air space of at least 1½" is necessary between the insulation and the roof sheathing
   D. no ventilation is needed

151

_____ 7. To improve the efficiency of louvers for ventilation in the end walls of gable roofs, _____.
   A. provide additional ventilation in the soffit area
   B. extend the attic insulation so it covers the soffit area
   C. install the louvers with the slats angled up
   D. place the louvers as far from the ridge as possible

_____ 8. A variation of a hip roof known as the Dutch hip is specifically designed to house _____ on both ends.
   A. globe type ventilators
   B. large triangular louvers
   C. continuous hip vents
   D. two rectangular louvers

_____ 9. The use of a ground cover in crawl spaces _____.
   A. is not recommended
   B. contributes to the decay of framing members
   C. eliminates the need for ventilators
   D. allows for the use of a smaller number of ventilators

_____ 10. The minimum free-air area for attic ventilators is based on the _____.
   A. total square footage of the home
   B. ceiling area of the rooms below
   C. attic floor area
   D. total roof surface

## Completion

Complete each sentence by inserting the correct answer on the line near the number.

_____ 1. If a vapor barrier ground cover is used in a crawl space, the total net area of crawl space ventilators should be _____ of the ground area.

_____ 2. _____ on ventilators should have as coarse a mesh as conditions permit.

_____ 3. The use of _____ prevents moisture from entering areas where condensation can occur.

_____ 4. The most commonly used material for a vapor barrier is _____.

_____ 5. Vapor barriers are not usually installed on _____, unless the temperatures commonly fall below –20°F or the roof slope is less than 3 in 12.

_____ 6. Materials of high vapor resistance should not be used on the _____ side of the wall.

_____ 7. The minimum free-air area of attic ventilators needed for a home with a ceiling area of 1500 square feet is _____ square feet.

_____  8. The minimum free-air area of ventilators needed for a crawl space that has a ground area of 3200 square feet is _____ square feet if a vapor barrier ground cover is used.

## Discussion

**Write your answer on the lines below the instruction.**

1. List the necessary steps that must be done during the construction process to eliminate condensation problems in a home.

   _____

   _____

   _____

Name _____ Date _____

# CHAPTER 51 CORNICE TERMS AND DESIGN

## Completion

**Complete each sentence by inserting the correct answer on the line near the number.**

_____ 1. Lookouts are framing members used to provide a nailing surface for the _____ of a wide cornice.

_____ 2. The soffit is an ideal location for the placement of _____.

_____ 3. The fascia is usually of ____ inch nominal thickness.

_____ 4. The portion of the fascia that extends below the soffit is called the _____ .

_____ 5. If the frieze is not used, the _____ is used to cover the joint between the siding and the soffit.

_____ 6. The most commonly used cornice design is the _____ cornice.

_____ 7. The cornice design that lacks a soffit is the _____ cornice.

_____ 8. The least attractive cornice design is the _____ cornice.

_____ 9. A rake cornice would be found on a home with a _____ or gambrel roof.

_____ 10. A _____ is constructed to change the direction of a level box cornice to the angle of the roof.

## Identification: Cornice Components

Identify each term, and write the letter of the correct answer on the line next to each number.

_____ 1. rafter

_____ 2. lookouts

_____ 3. soffit

_____ 4. fascia

_____ 5. frieze

_____ 6. cornice molding

_____ 7. plate

_____ 8. roof sheathing

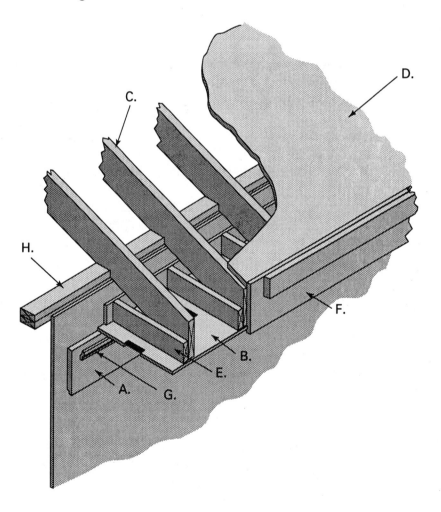

156

Name _____ Date _____

# CHAPTER 52  BUILDING CORNICES

## Multiple Choice

Write the letter for the correct answer on the line next to the number of the sentence.

_____ 1. Ends of each piece of finish lumber should not be _____.
   A. butted against each other
   B. stopped against each other
   C. left exposed
   D. sanded

_____ 2. A corner that is slightly rounded _____.
   A. holds a finish better than a sharp one
   B. is less dangerous than a sharp one
   C. is more attractive than a sharp one
   D. A, B, and C

_____ 3. The first member of the narrow box cornice to be installed is the _____.
   A. fascia
   B. soffit
   C. subfascia
   D. frieze

_____ 4. The outside edge of the soffit is slightly back beveled _____.
   A. to allow for ventilation
   B. to assure a tight fit between the soffit and fascia
   C. to make it easier to handle when installing
   D. to prevent it from splintering

_____ 5. When the ends of cornice finish are to be joined and only the face side shows, a _____ the ends assures a tight joint.
   A. slight standing cut on
   B. slight undercut on
   C. lightly rounding
   D. light coat of primer on

_____ 6. Nails used for fastening the soffit in place should be_____.
   A. set below the surface
   B. three times longer than the thickness of the soffit material
   C. galvanized common or casing nails
   D. A, B, and C

_____ 7. The fascia should extend below the soffit at least _____.
   A. 1"
   B. ¾"
   C. ½"
   D. ¼"

_____ 8. A wide box cornice differs from a narrow box cornice in that _____.
   A. the frieze is installed before the soffit
   B. lookouts must be installed to support the soffit
   C. a fascia is not necessary
   D. a subfascia is always needed

_____ 9. If the outside edge of a soffit fits into a groove on the fascia, _____.
   A. lookouts may be eliminated
   B. the soffit edge extends beyond the tail cut of the rafter
   C. a back bevel is necessary on the soffit
   D. A, B, and C

_____ 10. The soffit on a sloping box cornice is _____.
   A. eliminated
   B. attached to lookouts
   C. attached to the bottom edges of the rafter overhang
   D. installed when the roof is framed

## Completion

Complete each sentence by inserting the correct answer on the line near the number.

_____ 1. In a snub cornice, the rafter _____ is made flush with the outside edge of the wall framing.

_____ 2. In a snub cornice, the fascia also acts as the _____.

_____ 3. On an open cornice, the _____ must be cut between the rafters.

_____ 4. With an open cornice, _____ cuts on rafter tails must be absolutely straight.

_____ 5. With an open cornice, small metal circular louvers may be installed in the frieze between the _____ for ventilation.

_____ 6. The simplest type of rake trim is the _____ rake cornice.

_____ 7. The term _____ is given to any odd piece applied to fill or cover an opening or a defect.

_____ 8. On a snub rake cornice, the lower end of the rake fascia is usually trimmed after the _____ is installed.

_____ 9. With a moderate overhang on a box rake cornice, both the extending sheathing and a _____ rafter aid in supporting the rake cornice.

_____ 10. An extra wide overhang at the rake requires a _____ type frame for support of the rake cornice.

Name _____   Date _____

# CHAPTER 53  GUTTERS AND DOWNSPOUTS

## Completion

**Complete each sentence by inserting the correct answer on the line near the number.**

_____   1. The _____ carries water from the gutter downward and away from the foundation.

_____   2. No type of finish whatsoever is required on _____ gutters.

_____   3. For every _____ square feet of roof area, one square inch of gutter cross-section is needed.

_____   4. _____ gutters can be formed to practically any length when forming machines are brought to the job site.

_____   5. In order for water to drain toward the downspout, gutters should be installed with a pitch of about _____ inch(es) to 10 feet.

_____   6. On long buildings, the gutter is usually _____ in the center.

_____   7. A wood gutter should be _____ away from the fascia to allow air to circulate around it.

_____   8. When cutting a wood gutter, it should be placed _____ in the miter box to make the cut.

_____   9. _____ must be installed on a wood gutter wherever downspouts are to be located.

_____   10. To prevent _____ from occurring on metal gutters, be sure that the screws you are using are the same metal as the brackets.

_____   11. Aluminum brackets may be spaced up to _____ inches on center.

_____   12. _____ connectors are used to join the sections of metal or vinyl gutters.

_____   13. Round corrugated galvanized iron downspouts are fastened to the wall by galvanized iron rings called _____ .

_____   14. If downspouts are connected to the foundation, drain strainer caps must be placed over the gutter _____ .

159

## Identification: Gutter and Downspout Parts

Identify each term, and write the letter of the correct answer on the line next to each number.

_____ 1. end cap

_____ 2. slip connector

_____ 3. strap hanger

_____ 4. gutter

_____ 5. downspout

_____ 6. fascia bracket

_____ 7. spike and ferrule

_____ 8. conductor pipe band

_____ 9. end piece

_____ 10. strainer cap

_____ 11. elbow-style A

_____ 12. elbow-style B

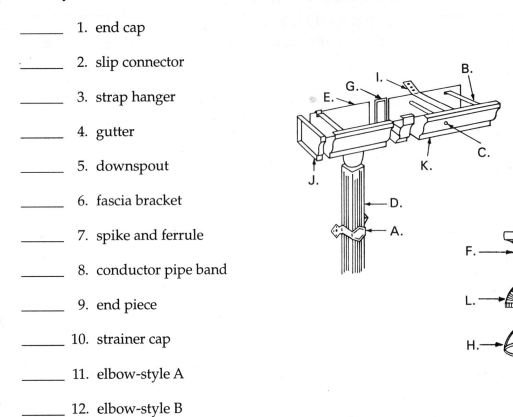

Name _____   Date _____

# CHAPTER 54  ASPHALT SHINGLES

## Multiple Choice

Write the letter for the correct answer on the line next to the number of the sentence.

_____ 1. When installing metal drip edge, _____ .
   A. tightly butt the end joints
   B. only use 1¼" roofing nails
   C. the roofing nails must be the same material as the drip edge
   D. space the nails every 12"

_____ 2. Asphalt shingle underlayment should _____ .
   A. be an entirely moisture-proof membrane
   B. allow the passage of water vapor
   C. be applied in vertical rows
   D. usually be as heavy a weight felt as is available

_____ 3. After the application of underlayment, it is recommended _____ .
   A. that metal drip edge be applied to the rakes
   B. that metal drip edge be applied to the eaves
   C. that the ridge vent be installed
   D. A, B, and C

_____ 4. Organic shingles have a base made of _____ .
   A. glass fibers
   B. heavy asphalt-saturated paper felt
   C. shredded thatching straw
   D. rubber

_____ 5. Mineral granules are used on the surface of asphalt shingles to _____ .
   A. provide weatherproofing qualities
   B. provide a good surface for asphalt cement to adhere to
   C. reduce its weight
   D. protect the shingle from the sun

_____ 6. _____ is generally determined by weight per square.
   A. Fire resistance
   B. Shingle quality
   C. Self-sealing ability
   D. Coverage

_____ 7. On long roofs, accurate vertical alignment is ensured by _____ .
   A. starting from either rake
   B. working right to left
   C. working left to right
   D. starting at the center and working both ways

161

_____ 8. It is recommended that no rake tab be less than _____ in width.
   A. 2"
   B. 3"
   C. 4"
   D. 5"

_____ 9. The purpose of the starter course is to _____.
   A. back up and fill in the spaces between the tabs on the first row of shingles
   B. eliminate the need of installing a drip edge
   C. reduce frost build up underneath the shingles
   D. A, B, and C

_____ 10. When fastening asphalt shingles, it is important to _____.
   A. use a minimum of three fasteners in each strip shingle
   B. follow the manufacturer's recommendations for application
   C. use a fastener long enough to penetrate the sheathing at least 3/8"
   D. not use power nailers

_____ 11. The most commonly used asphalt shingles have a maximum exposure of ___ inches.
   A. 3
   B. 5
   C. 7
   D. 9

_____ 12. When snapping a long chalk line, many times it is necessary to _____.
   A. first wet the line
   B. snap the line from the side closest to the chalk box
   C. hold the line against the roof with your thumb at about center and strike of the line
   D. simultaneously strike both sides of the line at the same time

_____ 13. When installing the ridge caps _____.
   A. coat the last two fasteners with asphalt cement
   B. start the shingles on the end away from the prevailing winds
   C. it may be necessary to warm them in cold weather
   D. A, B, and C

_____ 14. The maximum slope recommended for normal asphalt shingle application is _____ degrees.
   A. 45
   B. 50
   C. 55
   D. 60

## Matching

Write the letter for the correct answer on the line near the number to which it corresponds.

_____ 1. square

_____ 2. electrolysis

_____ 3. end lap

_____ 4. deck

_____ 5. course

_____ 6. flashing

_____ 7. asphalt cements

_____ 8. top or head lap

_____ 9. exposure

A. horizontal rows of shingles or roofing

B. strips of thin sheet metal used to make watertight joints

C. trowel applied adhesives used to bond asphalt roofing products

D. the amount of roofing required to cover 100 square feet

E. a reaction that occurs between unlike metals when wet

F. horizontal distance the ends of roofing in the same course overlap

G. the amount of roofing in each course subjected to the weather

H. the wood roof surface to which roofing is applied

I. shingle height minus the exposure

Name _____  Date _____

# CHAPTER 55   ROLL ROOFING

## Completion

Complete each sentence by inserting the correct answer on the line near the number.

_____  1. Roll roofing can be installed on roofs that slope as little as _____ inch of rise per foot of run.

_____  2. A concealed nail type of rolled roofing called _____ has a top lap of 19 inches.

_____  3. All kinds of rolled roofing come in rolls that are _____ inches wide.

_____  4. When roofs have a pitch that is less than _____ inches rise per foot, it is recommended that rolled roofing be used.

_____  5. On pitches that are less than 2 inches of rise per foot, the _____ type of roofing should not be used.

_____  6. The same type and length of _____ as are used on asphalt shingles should be used on rolled roofing.

_____  7. Rolled roofing's coat can crack if it is applied at temperatures below _____ degrees Fahrenheit.

_____  8. Use only the lap or quick setting cement recommended by the _____ .

_____  9. Strips of rolled roofing 9" wide should be applied along the eaves and rakes with about a _____ inch overhang.

_____  10. All rolled roofing end laps should be _____ inches wide and cement should be applied the full width of the lap.

## Discussion

Write your answer on the lines below the instruction.

1. Due to the fact that rolled roofing is made by several different manufacturers, the builder must be aware of what things when applying it?

_____

_____

_____

Name _____  Date _____

# CHAPTER 56 WOOD SHINGLES AND SHAKES

## Multiple Choice

Write the letter for the correct answer on the line next to the number of the sentence.

_____ 1. Most shingles and shakes are produced from _____.
   A. white pine
   B. poplar
   C. western red cedar
   D. redwood

_____ 2. Wood shingles have a _____ surface.
   A. somewhat rough
   B. relatively smooth sawn
   C. highly textured, natural grain, split
   D. very smooth

_____ 3. There are _____ standard grades of wood shingles.
   A. 2
   B. 3
   C. 4
   D. 5

_____ 4. The area covered by one square of shingles or shakes depends on the _____.
   A. amount of shingle or shake exposed to the weather
   B. weight of the square
   C. choice of underlayment
   D. choice of deck material

_____ 5. Shingles and shakes may be applied _____
   A. over spaced or solid roof sheathing
   B. on roofs with slopes under 3" of rise per foot
   C. only if extra heavy underlayment is used
   D. A, B, and C

_____ 6. The sliding gauge on a shingling hatchet is used _____.
   A. to split shakes
   B. for checking the shingle exposure
   C. to extract fasteners
   D. to trim shingles and shakes

_____ 7. The use of a power nailer on wood shingles and shakes _____.
   A. is a tremendous time saver
   B. is not permitted
   C. could result in more time lost than gained
   D. lengthens the life expectancy of the roof

167

_____ 8. _____ nails are corrosion-resistant.
   A. Stainless steel
   B. Hot-dipped galvanized
   C. Aluminum
   D. A, B, and C

_____ 9. If staples are used to fasten wood shingles or shakes, they should be _____.
   A. at least 20 gauge with a 3/8" minimum crown
   B. long enough to penetrate the sheathing by at least 1/4"
   C. driven with the crown across the grain
   D. long enough to penetrate the sheathing by at least 3/8"

_____ 10. If a gutter is used, overhang the wood shingle starter course _____.
   A. 1/2" past the fascia
   B. plumb with the center of the gutter
   C. 1" past the inside edge of the gutter
   D. 1 1/2" past the fascia

## Completion

**Complete each sentence by inserting the correct answer on the line near the number.**

_____ 1. Place each fastener about _____ inch in from the edge of the wood shingle and not more than 1" above the exposure line.

_____ 2. Do not allow the head of the fastener to _____ the surface of the shingle.

_____ 3. In regions of heavy snowfall, it is recommended that the starter course be _____.

_____ 4. Joints in adjacent courses of wood shingles should be staggered at least _____ inches.

_____ 5. No joint in any _____ adjacent courses should be in alignment.

_____ 6. On intersecting roofs, do not break joints in the _____.

_____ _____ 7. Hip and ridge caps are usually ____ to ____ inches wide.

_____ 8. When laying wood shakes, an _____ consisting of strips of 18" wide, #30 roofing felt is used.

_____ 9. Straight-split shakes should be laid with their _____ end toward the ridge.

_____ 10. It is important to regularly check the _____ of wood shakes with the hatchet handle since there is a tendency for the course to angle toward the ground.

_____ 11. Divide the total square feet of the roof area by _____ to determine the total number of squares needed for the job.

_____ 12. At standard exposure, estimate ____ pounds of nails per square of shingles.

Name _____    Date _____

# CHAPTER 57  FLASHING

## Multiple Choice

Write the letter for the correct answer on the line next to the number of the sentence.

_____ 1. The woven valley method of flashing _____ .
   A. is an open valley method
   B. is a closed valley method
   C. should have the course end joints occurring on the valley center line
   D. requires sheet metal flashing must be used under it

_____ 2. The closed cut method of valley flashing _____ .
   A. requires the use of fasteners at the valley's center
   B. does not require the use of 50 pound per square rolled roofing
   C. is an open valley method
   D. requires the use of asphalt cement

_____ 3. When using step flashing in a valley, each piece of flashing should be at least _____ wide, if the roof has less than a 6" rise.
   A. 6"
   B. 12"
   C. 18"
   D. 24"

_____ 4. The height of each piece of step valley flashing should be at least 3" more than the shingle _____ .
   A. head lap
   B. top lap
   C. exposure
   D. salvage

_____ 5. When the valley is completely flashed with the step flashing method, _____ .
   A. a 6" wide strip of metal flashing the length of the valley is visible
   B. rolled roofing is applied over the top of it
   C. no metal flashing surface is exposed
   D. a coating of asphalt cement is applied on top of the shingles

_____ 6. The usual method of making the joint between a vertical wall and a roof watertight is the use of _____ .
   A. a lead saddle
   B. an apron
   C. a heavy coating of asphalt cement
   D. step flashing

_____ 7. On steep roofs between the upper side of the chimney and the roof deck, a saddle or _____ is built.
   A. heel
   B. frog
   C. cricket
   D. gusset

_____ 8. Chimney flashings are usually installed by _____.
   A. carpenters
   B. brick masons
   C. roofers
   D. laborers

_____ 9. The upper ends of chimney flashing is _____.
   A. fastened to the chimney with concrete nails driven through the flashing and into the mortar joints
   B. imbedded in asphalt cement against the chimney
   C. bent around and mortared in between the courses of brick
   D. bent over and bedded to the shingles with asphalt cement

_____ 10. When installing flashing over a stack vent, _____.
   A. shingle over top of the lower end of the stack vent flashing
   B. the flashing is attached with one fastener in each upper corner
   C. be sure the shingle fasteners penetrate the flashing
   D. A, B, and C

## Completion

**Complete each sentence by inserting the correct answer on the line near the number.**

_____ 1. _____ and zinc are expensive but high quality flashing material.

_____ 2. When reroofing, it is a good practice to replace all the _____.

_____ 3. It is necessary to install an eaves flashing whenever there is a possibility of _____ forming.

_____ 4. Roof _____ are especially vulnerable to leaking due to the great volume of water that flows down through them.

_____ 5. When using rolled roofing as valley flashing, the first layer should be laid with its mineral surface side _____.

_____ 6. When the end courses are properly trimmed on a valley that is 16' long, its width at the ridge will be 6" and at the eaves _____ inches.

_____ 7. The metal flashing in an open valley between a low-pitched roof and a much steeper one, should have a _____ inch high crimped standing seam in its center.

_____  8. _____ valleys are those where the shingles meet in the center of the valley covering the valley flashing completely.

_____  9. The rolled roofing required in closed valleys must be _____ pound per square or more.

_____  10. On woven valleys, be sure that no _____ is located within 6" of the centerline.

## Discussion

**Write your answer on the lines below the instruction.**

1. Perhaps the most important thing to keep in mind when applying any roofing material is that water always runs downhill. Write a short paragraph explaining the importance of this statement to the application of roofing materials.

   _____

   _____

   _____

Name _____   Date _____

# CHAPTER 58  WINDOW TERMS AND TYPES

## Multiple Choice

Write the letter for the correct answer on the line next to the number of the sentence.

_____ 1. Most windows are primed _____ with their first coat of paint.
   A. after installation
   B. at the factory
   C. by the retailer
   D. when the siding is applied

_____ 2. Vinyl-clad wood windows _____ .
   A. come in a large variety of colors
   B. are designed to eliminate painting
   C. must be primed before installation
   D. are only available in fixed windows

_____ 3. Screens are attached to the outside of the window frame on _____ windows.
   A. awning
   B. sliding
   C. double-hung
   D. jalousie

_____ 4. _____ fixed windows are widely used in combination with other window types.
   A. elliptical
   B. half rounds
   C. quarter rounds
   D. A, B, and C

_____ 5. The single-hung window is similar to the double-hung window except _____ .
   A. the lower sash swings outward
   B. the parting bead is eliminated
   C. the upper sash is fixed
   D. it only has one sash

_____ 6. When the sashes are closed on double-hung windows, specially shaped _____ come together to form a weather-tight joint.
   A. parting beads
   B. meeting rails
   C. blind stops
   D. sash locks

_____ 7. The _____ window consists of a sash hinged at the side and swinging outward by means of a crank or lever.
   A. casement
   B. awning
   C. hopper
   D. jalousie

_____ 8. An advantage of the casement type window is that _____.
   A. the entire sash can be opened for maximum ventilation
   B. its system of springs and balances makes it easy to open
   C. they come with removable sashes for easy cleaning
   D. the screens are installed on the outside of the frame

_____ 9. An awning window consists of a frame in which a sash _____.
   A. hinged at the bottom swings inward
   B. slides horizontally left or right in a set of tracks
   C. hinged at the side and swings inward
   D. hinged at the top swings outward

_____ 10. A major disadvantage to jalousie windows is that _____.
   A. ventilation is poor with this window type
   B. they are not very energy efficient
   C. they are not recommended in warmer climates
   D. screens cannot be used with them

## Completion

Complete each sentence by inserting the correct answer on the line near the number.

_____ 1. Wood windows, doors, and cabinets fabricated in woodworking plants are referred to as _____.

_____ 2. The _____ is the frame that holds the glass in the window.

_____ 3. _____ are the vertical edge members of the sash.

_____ 4. Small strips of wood that divide the glass into smaller lights are called _____.

_____ 5. Many windows come with false muntin called _____.

_____ 6. The installation of glass in a window sash is called _____.

_____ 7. Skylights and roof windows are generally required to be glazed with _____.

_____ 8. To raise the R-value of insulating glass, the space between the glass is filled with _____ gas.

_____ 9. An invisible, thin _____ coating is bonded to the air space side of the inner glass of solar control insulating glass.

_____ 10. The bottom horizontal member of the window frame is called a _____.

_____ 11. The _____ are the vertical sides of the window frame.

_____ 12. A _____ is formed where the two side jambs on side by side windows are joined together.

## Identification: Windows

Identify each term, and write the letter of the correct answer on the line next to each number.

_____ 1. top rail

_____ 2. bottom rail

_____ 3. stile

_____ 4. muntin

_____ 5. light

_____ 6. double-hung window

_____ 7. casement window

_____ 8. awning window

_____ 9. sliding windows

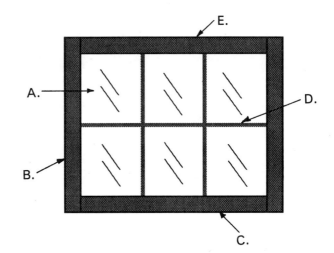
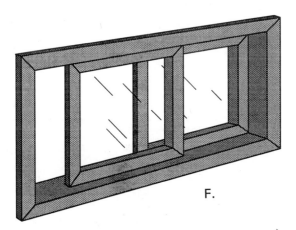
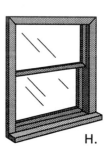
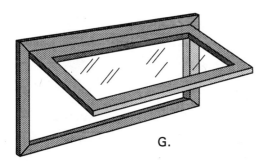

175

Name _____    Date _____

# CHAPTER 59  WINDOW INSTALLATION AND GLAZING

## Multiple Choice

Write the letter for the correct answer on the line next to the number of the sentence.

_____ 1. Those responsible for planning the location or selection of windows need to be aware of building code requirements for minimum _____ .
   A. areas of natural light
   B. ventilation by windows
   C. window size in regard to emergency egress
   D. A, B, and C

_____ 2. _____ windows should not be located above porches or decks unless they are high enough to allow people to travel under them.
   A. double- and single-hung
   B. sliding and hopper
   C. awning and casement
   D. fixed

_____ 3. The builder should refer to the window _____ to determine the window style, size, manufacturer's name, and unit number.
   A. schedule
   B. agenda
   C. program
   D. plan

_____ 4. In order for the builder to better understand the construction of a particular window unit, he should refer to the _____ .
   A. local building codes
   B. Sweets register
   C. the manufacturer's catalog
   D. window syllabus

_____ 5. When applying building paper or a housewrap prior to siding application, be sure it is one that _____ .
   A. is completely moisture-tight
   B. will permit the passage of water vapor to the outside
   C. will allow infiltration of fresh air through its surface
   D. will permit moisture passage to the building's interior

_____ 6. _____ will survive the longest period of exposure to the weather.
   A. Building paper
   B. House wrap
   C. Polyethylene film
   D. Asphalt felt

177

_____ 7. If building paper is to be used, it must be applied _____ .
   A. prior to the installation of the doors and windows
   B. immediately after framing is completed
   C. after the doors and windows are installed
   D. after the insulation is installed

_____ 8. When using housewrap overlap all joints by at least _____ inch(es).
   A. 1
   B. 3
   C. 6
   D. 9

_____ 9. Most windows are installed so that _____ .
   A. their tops are all at different levels
   B. their header casings lie at the same elevation
   C. the same size of window is on each story
   D. the double-hung windows are on the first floor and the single-hung windows are on the second floor

_____ 10. When installing windows, _____ .
   A. remove any diagonal braces applied at the factory
   B. unlock and open the sash
   C. leave all protection blocks on the the window unit
   D. be sure to shim level and plumb the unit

## Completion

**Complete each sentence by inserting the correct answer on the line near the number.**

_____ 1. When it is necessary to shim the bottom of a window to level it, the shim is placed between the rough sill and the bottom end of the window's side _____ .

_____ 2. To avoid splitting the casing, all nails should be at least _____ inches back from casing's end.

_____ 3. Nails driven through the window casing should be _____ casing or common nails.

_____ 4. On vinyl-clad windows, large head _____ nails are driven through the nailing phalange instead of the casing.

_____ 5. Windows are installed in masonry openings against wood _____ .

_____ 6. Trades people who perform the work of cutting and installing lights of glass in sash or doors are known as _____ .

_____ 7. _____ are small triangular or diamond-shaped pieces of thin metal used to hold glass in place.

_____ 8. A light of glass is installed with its crown side _____ in a thin bed of compound against the rabbet of the opening.

_____ 9. Going over the scored line a second time with the glass cutter will _____ the glass cutter.

_____ 10. Prior to scoring a line on glass, it is recommended to brush some _____ along the line of cut.

Name _____ Date _____

# CHAPTER 60 DOOR FRAME CONSTRUCTION AND INSTALLATION

## Completion

Complete each sentence by inserting the correct answer on the line near the number.

_____ 1. Exterior doors, like windows, are manufactured in _____ plants in a wide variety of styles and sizes.

_____ 2. Many entrance doors come _____ in frames, with complete exterior casings applied, ready for installation.

_____ 3. The bottom member of an exterior door is the _____ .

_____ 4. Vertical side members of an exterior door are known as _____ .

_____ 5. In residential construction, exterior doors usually swing _____ .

_____ 6. In order to ensure an exact fit between the sill and the door, sometimes the _____ at the bottom of the door is adjustable.

_____ 7. For walls of odd thicknesses, _____ may be ripped to any desired width, unless they are double-rabbeted.

_____ 8. To construct an accurately sized exterior door frame, it is advisable to have the door available for _____ .

_____ 9. Jamb _____ is equal to the overall wall thickness from the outside of the wall sheathing to the inside surface of the interior wall covering.

_____ 10. If jamb stock must be ripped, rip the edge opposite the _____ .

_____ 11. Side jambs are usually _____ to receive the ends of the header jamb and the sill.

_____ 12. On each side of the door, a _____ inch joint should be allowed between the door and the frame.

_____ 13. Drive a _____ between the back sides of both the sill and header jamb and the shoulders of the dado, before nailing to assure a tight fit on the face side.

_____ 14. The _____ is the amount of setback of the casing from the inside face of the door jamb.

_____ 15. When square-edged casings are used, a _____ makes a weather-tight joint between them.

**181**

_____ 16. When an exterior molded casing is subjected to severe weather, a _____ miter joint is used.

_____ 17. When setting a door frame, cut off the _____ which project beyond the sill and header jamb.

_____ 18. If necessary to level the sill when setting the door frame, the shims are placed under the _____ .

_____ 19. A _____ is a twist in the door frame caused when the side jambs do not line up vertically with each other.

## Identification: Exterior Door Components

Identify each term, and write the letter of the correct answer on the line next to each number.

_____ 1. threshold

_____ 2. head casing

_____ 3. side casing

_____ 4. head jamb

_____ 5. sill

_____ 6. band mold

_____ 7. side jamb

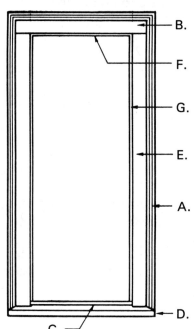

## Discussion

Write your answer on the lines below the instruction.

1. In buildings used by the general public that have an occupancy load that exceeds fifty, building codes require exterior doors used as exits to swing outward. Why is this so?

_____

_____

_____

Name _____  Date _____

# CHAPTER 61   DOOR FITTING AND HANGING

## Multiple Choice

**Write the letter for the correct answer on the line next to the number of the sentence.**

_____ 1. _____ doors are highly crafted designer doors with a variety of cut-glass designs.
   A. French
   B. Dutch
   C. High-style
   D. Sash

_____ 2. _____ doors consist of top and bottom units, hinged independently of each other.
   A. Dutch
   B. Flush
   C. French
   D. Ventilating

_____ 3. A panel door consists of a frame that surrounds panels of _____ .
   A. solid wood
   B. glass
   C. louvers
   D. A, B, and C

_____ 4. The outside vertical members of a panel door are called _____ .
   A. mullions
   B. rails
   C. stiles
   D. jambs

_____ 5. The widest of all rails in a panel door is the _____ rail.
   A. bottom
   B. top
   C. lock
   D. intermediate

_____ 6. Practically all exterior entrance doors are manufactured with a thickness of ____ inches.
   A. ¾
   B. 1
   C. 1½
   D. 1¾

_____ 7. Prehung exterior doors come _____ .
   A. already fixed and hinged in the door frame
   B. with the lockset installed
   C. the outside casing installed
   D. A, B, and C

_____ 8. When installing a prehung door unit, _____.
   A. remove the factory installed spacers between the door and the jamb before leveling and shimmying
   B. and nailing the side jambs to the studs, avoid nailing through the shims
   C. make sure the outside edge of the jamb is flush with the wall unit
   D. it is good practice to hammer the nails completely flush with the finish

_____ 9. The first step in fitting a door is to _____.
   A. determine the side that will close against the stops on the door frame
   B. install the lockset
   C. apply the hinges
   D. attach ¼" spacers to the door's stiles

_____ 10. Exterior doors containing lights of glass must be hung with the _____.
   A. removable glass bead facing the exterior
   B. glass bead first removed
   C. removable glass bead facing the interior
   D. lights of glass removed

## Completion

Complete each sentence by inserting the correct answer on the line near the number.

_____ 1. When a sash door is manufactured with _____ glazing, the possibility of any water seeping through the joints is virtually eliminated.

_____ 2. The _____ of the door is designated as being either left-hand or right-hand.

_____ 3. A left-handed door has its hinges on the left side if you are standing on the side that swings _____ from you.

_____ 4. The process of fitting a door into a frame is called _____.

_____ 5. The lock edge of a door must be planned on a _____.

_____ 6. When fitting doors, extreme care must be taken not to get them _____.

_____ 7. To _____ sharp corners means to round them over slightly.

_____ 8. Use _____ 4" × 4" hinges on 1¾" doors 7'-0" or less in height.

_____ 9. The recess in a door for the hinge is called the _____ or sometimes, a hinge mortise.

_____ 10. When laying out hinges, use a _____ instead of a pencil to mark the line.

_____ 11. When many doors need to be hung, a butt hinge template and a portable electric _____ is used, cut hinge gains.

_____ 12. An _____ is a molding that is rabbeted on both edges and designed to cover the joint between double doors.

Name _____    Date _____

# CHAPTER 62  DOOR LOCK INSTALLATION

## Completion

**Complete each sentence by inserting the correct answer on the line near the number.**

_____  1. Key-in-knob is another name for _____ locksets.

_____  2. A distinguishing difference between the two major categories of locksets are their basic _____ .

_____  3. An _____ lock combines deadbolt locking action with a standard key-in-knob set.

_____  4. _____ locks are not usually used in residential construction because of their high cost, and the increased amount of time installation takes.

_____  5. _____ are decorative plates of various shapes that are installed between the lock handle or knob.

_____  6. Measuring up from the floor, the recommended distance to the centerline of a lock is usually _____ inches.

_____  7. The _____ of the lock is the distance from the edge of the door to the center of the hole through the side of the door.

_____  8. When boring the holes for a lockset, the hole through the _____ of the door should be bored first.

_____  9. The striker plate is installed on the door _____ .

_____  10. After the door is fitted, hung, and locked, remove all _____ and prime the door and all exposed parts of the door frame.

## Discussion

**Write your answer on the lines below the instruction.**

1. When laying out hinges, face plates, and striker plates, it is recommended that a sharp knife be used instead of a pencil. Why is this so?

   _____

   _____

   _____

185

Name _____   Date _____

# CHAPTER 63  WOOD SIDING TYPES AND SIZES

## Multiple Choice

Write the letter for the correct answer on the line next to the number of the sentence.

_____ 1. Most redwood siding is produced by mills that belong to _____ .
   A. WWPA
   B. CRA
   C. APA
   D. AHWA

_____ 2. In sidings classified as _____ , the annual growth rings must form an angle of 45° or more with the surface.
   A. flat grain
   B. vertical grain
   C. mixed grain
   D. cross grain

_____ 3. Vertical grain siding is the highest quality because _____ .
   A. it warps less
   B. takes and holds finishes better
   C. has less defects and is easier to work
   D. A, B, and C

_____ 4. _____ surfaces generally hold finishes longer than other types of surfaces.
   A. Smooth
   B. Flat-grained
   C. Saw-textured
   D. Double-planed

_____ 5. Knotty grade siding is divided into #1, #2, and #3 common depending on _____ .
   A. the type and number of knots
   B. its approved method of application
   C. the species of tree it comes from
   D. the thickness it is cut to

_____ 6. The best grades of redwood siding are grouped in a category called _____ .
   A. structural
   B. edifical
   C. clear
   D. architectural

_____ 7. Bevel siding is more commonly known as _____ .
   A. drop
   B. tongue and groove
   C. clapboard
   D. channel rustic

187

_____ 8. Most panel and lap siding is manufactured from _____.
   A. plywood and hardboard
   B. cedar and redwood
   C. hemlock and poplar
   D. A, B, and C

_____ 9. Most panel siding is shaped with _____ edges for weathertight joints.
   A. back beveled
   B. chamfered
   C. shiplapped
   D. mitered

_____ 10. Lap siding comes in thicknesses from 7/16 to 9/16 of an inch, widths of 6, 8, and 12 inches and lengths of _____ feet.
   A. 4
   B. 8
   C. 12
   D. 16

## Math

Without subtracting for window and door openings, estimate how many square feet of area is to be covered with siding on the house shown below.

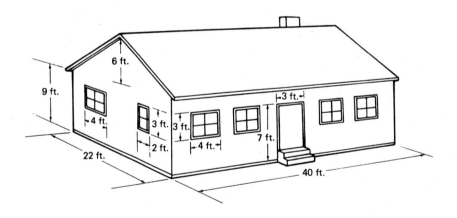

Name _____   Date _____

# CHAPTER 64  APPLYING HORIZONTAL AND VERTICAL WOOD SIDING

## Multiple Choice

Write the letter for the correct answer on the line next to the number of the sentence.

_____ 1. Corner boards are not usually used when wood siding is applied _____.
   A. vertically
   B. horizontally
   C. diagonally
   D. A, B, and C

_____ 2. When applying tongue and groove siding vertically, fasten by _____.
   A. face nailing into the groove edge of each piece
   B. toenailing into the groove edge of each piece
   C. toenailing into the tongue edge of each piece
   D. face nail through groove edge and into the tongue edge of the next piece

_____ 3. If necessary to make horizontal joints between lengths of vertical siding, _____.
   A. a butt joint alone is acceptable
   B. install building paper beneath the joint
   C. use mitered or rabbeted end joints
   D. leave a 1/16" gap to receive caulking

_____ 4. When applying short lengths of vertical siding under a window, _____.
   A. leave an ample margin for expansion of the window
   B. apply a bead of exterior glue between the top of the siding and the window's bottom
   C. rabbet the top of the siding to fit into the weather groove on the window's bottom
   D. butt the pieces against the window's bottom

_____ 5. The last piece of vertical siding should be _____.
   A. much narrower than the rest
   B. wider than the rest
   C. fastened with screws instead of nails
   D. as close as possible in width to the pieces previously installed

_____ 6. When installing vertical panel siding that is thicker than ½", use _____ siding nails.
   A. 4d
   B. 6d
   C. 8d
   D. 10d

_____ 7. When installing horizontal or vertical panel siding, it is important that all horizontal joints be _____.
   A. offset and lapped
   B. rabbeted
   C. flashed if a butt joint
   D. A, B, and C

189

_____ 8. Metal or plastic corners are many times used on siding made up of composition material because _____.
   A. composition material does not miter well
   B. corner boards are not permitted
   C. composition material weathers poorly
   D. A, B, and C

## Completion

Complete each sentence by inserting the correct answer on the line near the number.

_____ 1. A _____ is finish work that may be installed around the perimeter of the building slightly below the top of the foundation.

_____ 2. If the siding terminates against the soffit and no frieze is used, the joint between them is covered by a _____ molding.

_____ 3. If neither corner boards nor metal corners are used, then horizontal siding may be _____ around exterior corners.

_____ 4. On interior corners, the siding courses may butt against a square corner _____.

_____ 5. One of the two pieces making up an outside corner board should be narrower than the other by the thickness the _____.

_____ 6. Before installing corner boards, _____ both sides of the corner with #15 felt, applied vertically on each side.

_____ 7. A major advantage of bevel siding over other types is the ability to vary the siding's _____.

_____ 8. The furring strip, which is applied before the first course of siding, must be the same thickness and width as the siding _____.

_____ 9. A _____ is often used for accurate layout of siding where it butts against corner boards, casings and similar trim.

_____ 10. For weather-tightness when fitting siding under a window, it is important that siding fits snugly in the _____ on the underside of the window sill.

_____ 11. Bevel and Dolly Varden siding are only to be applied in the _____ position.

_____ 12. When installing vertical tongue and groove siding directly to the frame, _____ must be provided between the studs.

Name _____   Date _____

# CHAPTER 65  WOOD SHINGLE AND SHAKE SIDING

## Completion

**Complete each sentence by inserting the correct answer on the line near the number.**

_____ 1. Rebutted and rejointed machine-grooved sidewall shakes have _____ faces.

_____ 2. Fancy butt shingles were widely used in the 19th century on _____ style buildings.

_____ 3. Red cedar shingles are available in several styles and exposures on factory applied 4- and 8- foot _____ .

_____ 4. When applying shingles to _____ , greater exposures are permitted than on roofs.

_____ 5. When more than one layer of shingles is needed, less expensive _____ shingles are used for the underlayers.

_____ 6. Untreated shingles should be spaced ⅛" to ¼" apart to allow for _____ and to prevent buckling.

_____ 7. Use a _____ when it is necessary to trim and fit the edges of wooden shingles.

_____ 8. If rebutted and rejointed shingles are used, no _____ should be necessary.

_____ 9. Shingles should be fastened with two nails or staples about _____ in from the edge.

_____ 10. Fasteners used on shingles, should be hot-dipped galvanized, stainless steel, or aluminum and driven about 1" above the butt line of the next _____ .

_____ 11. Shingles in panelized form are applied in the same manner as horizontal _____ .

_____ 12. On outside corners, shingles may be applied by alternately _____ each course in the same manner as applying a wood shingle ridge.

_____ 13. When double coursing wood shingles, the first course is _____ .

_____ 14. The number of squares of shingles needed to cover a certain area depends on how much of them are _____ to the weather.

_____ 15. One square of shingles will cover 100 square feet when 16" shingles are exposed _____ inches.

191

## Identification: Fancy Butt Shingles

Identify each term, and write the letter of the correct answer on the line next to each number.

_____ 1. arrow

_____ 2. round

_____ 3. diagonal

_____ 4. octagonal

_____ 5. square

_____ 6. diamond

_____ 7. hexagonal

_____ 8. fish scale

_____ 9. half cove

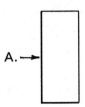

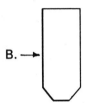
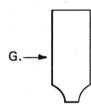
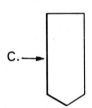
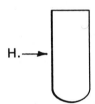
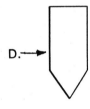

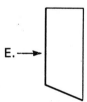

Name _____   Date _____

# CHAPTER 66 ALUMINUM AND VINYL SIDING

## Multiple Choice

Write the letter for the correct answer on the line next to the number of the sentence.

_____ 1. Aluminum and vinyl siding systems are _____ .
   A. both finished with baked on enamel
   B. similar to each other except for the material
   C. expansion resistant
   D. only available for horizontal applications

_____ 2. Aluminum and vinyl siding panels for horizontal applications are _____ .
   A. made in 8″ and 12″ widths
   B. also used for soffits
   C. made in 6″ and 9″ widths
   D. made in configurations to resemble 4, 5, or 6 courses of Dolly Varden siding

_____ 3. Panels designed for vertical application, come in _____ widths and are shaped to resemble boards.
   A. 6″
   B. 8″
   C. 10″
   D. 12″

_____ 4. With changes in temperature, siding may contract and expand as much as _____ in a 12′-6″ section.
   A. 1/8″
   B. 1/4″
   C. 3/8″
   D. 1/2″

_____ 5. Fasteners should be driven _____ the siding.
   A. very tightly against
   B. not too tightly into
   C. every 24″ along
   D. to the left of center in the slots on

_____ 6. Starter strips must be applied _____ .
   A. tightly against corner posts
   B. at vertical intervals of every four feet
   C. as straight as possible
   D. only on siding that is installed vertically

**193**

_____ 7. Install _____ across the tops and along the sides of window and door casings.
   A. starter strips
   B. corner posts
   C. undersill
   D. J-channel

_____ 8. When marking the cutout of a siding panel under a window, allow for _____ clearance under and on either side of the window.
   A. 1/16"
   B. 1/8"
   C. 1/4"
   D. 3/8"

_____ 9. _____ applied on the wall up against the soffit, prior to the installation of the last course of siding panel.
   A. Starter strips are
   B. Corner posts are
   C. Undersill trim is
   D. J-channel is

_____ 10. The layout for vertical siding should be planned so that _____.
   A. the same panel width is exposed at both ends of the wall
   B. any panels that need to have their edges cut to size are placed in the rear of the building
   C. any width adjustment made in the panels come in the center of the wall
   D. no panels are to have their edges trimmed

# Identification: Vinyl Siding Systems

Identify each term, and write the letter of the correct answer on the line next to each number.

_____ 1. horizontal siding starter strip

_____ 2. vertical siding or soffit

_____ 3. fascia

_____ 4. undersill trim

_____ 5. outside corner post

_____ 6. J-channel

_____ 7. inside corner post

_____ 8. perforated soffit

_____ 9. undersill finish trim

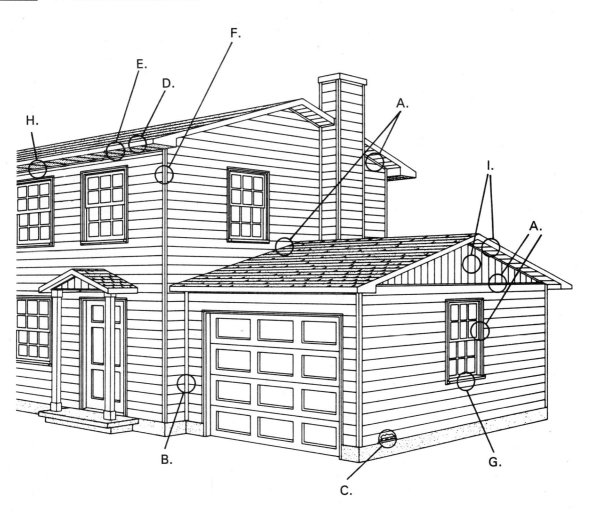

195

Name _____   Date _____

# CHAPTER 67  PORCH AND DECK CONSTRUCTION

## Multiple Choice

Write the letter for the correct answer on the line next to the number of the sentence.

_____ 1. Girders are installed on the beams using _____ .
   A. post and beam metal connectors
   B. fiber tube forms
   C. post anchors
   D. joist hangers

_____ 2. When joists are hung between the girders, _____ .
   A. joist hangers are not necessary
   B. the overall length of the deck must be decreased
   C. the overall depth of the deck is decreased
   D. they must be installed with the crown down

_____ 3. When decking is run _____ to the joists, the spacing of the joists may be 24" on center.
   A. at right angles
   B. parallel
   C. diagonal
   D. A, B, and C

_____ 4. If the deck is less than _____ above the ground, the supporting posts will not need to be permanently braced.
   A. 4'
   B. 6'
   C. 8'
   D. 10'

_____ 5. When applying the deck boards, _____ .
   A. lay them with the bark side down
   B. it is advisable to start at the inside edge
   C. maintain about a ½" between them
   D. make tight fitting end joints and stagger them between adjacent rows

_____ 6. A _____ board may be fastened around the perimeter of the deck with its top edge flush with the top of the deck.
   A. ledger
   B. fascia
   C. batter
   D. skirt

197

_____ 7. If the deck is more than 30" above the ground, most codes require _____ high railing around the exposed sides.
   A. 24"
   B. 28"
   C. 32"
   D. 36"

_____ 8. _____ are another name for railing posts.
   A. Stanchions
   B. Balusters
   C. Preachers
   D. Rails

_____ 9. The space between the top and bottom rail may be filled with _____.
   A. intermediate rails
   B. ballisters
   C. lattice work
   D. A, B, or C

_____ 10. If benches are built into the deck, their seat height should be _____ above the deck.
   A. 14"
   B. 16"
   C. 18"
   D. 20"

## Completion

**Complete each sentence by inserting the correct answer on the line near the number.**

_____ 1. Lumber used in deck construction must be either from a decay resistant species or be _____.

_____ 2. It is the _____ of redwood and cedar that is resistant to decay.

_____ 3. If pressure-treated southern pine is used in deck construction, the grade of _____ is structurally adequate for most applications.

_____ 4. _____ is the most suitable and economical grade of California redwood for deck post, beams, and joists.

_____ 5. All nails, fasteners, and hardware should be stainless steel aluminum or _____ galvanized.

_____ 6. When a deck is constructed against a building, a _____ is nailed or bolted against the wall for the entire length of the deck.

_____ 7. After the deck is applied, a _____ is installed under the siding and on top of the deck board.

_____ 8. All deck footings require digging a hole and filling it with _____.

_____ 9. In cold climates all deck footings must extend below the _____.

_____ 10. All supporting posts are set on footings then braced _____ in both directions.

Name _____   Date _____

# CHAPTER 68   FENCE DESIGN AND ERECTION

## Multiple Choice

Write the letter for the correct answer on the line next to the number of the sentence.

_____ 1. For the strongest fences, set the posts in _____ .
   A. clay
   B. gravel
   C. concrete
   D. mortar

_____ 2. Filling the bottom of the fence post hole with gravel _____ .
   A. reduces the required depth of the hole
   B. strengthens the post
   C. eliminates the need to brace the post
   D. helps extend the life of the post

_____ 3. When placing concrete around fence posts, _____ .
   A. form a slight depression around the post
   B. form the top so it pitches down from the post
   C. tamp it level
   D. use a very wet mix that has been allowed to partially harden

_____ 4. An alternative to embedding the fence posts in concrete is to _____ .
   A. pack gravel around the post
   B. attach the post to a metal anchor embedded in the concrete
   C. pack the post in sand
   D. A, B, or C

_____ 5. When installing rails on fences, keep the bottom rail at least _____ above the ground.
   A. 2"
   B. 4"
   C. 6"
   D. 8"

_____ 6. When the situation exists where the faces of the wooden posts are not in the same line as the rails, then the _____ .
   A. posts must be reset
   B. rail ends must be cut out of square to match the posts
   C. rail ends must be perfectly square
   D. rails must be slightly bowed to align them

_____ 7. If iron posts are boxed with wood, then the rails are installed _____ .
   A. in the same manner as for wood posts
   B. with special metal pipe grips
   C. by boring holes in the rails and sliding them over the posts
   D. with porcelain insulators

199

_____ 8. When applying spaced pickets, _____.
   A. use a picket or a ripped piece of lumber for a spacer
   B. cut only the bottom end of the pickets when trimming their height
   C. the bottom of the pickets should not touch the ground when installed
   D. A, B, and C

## Completion

**Complete each sentence by inserting the correct answer on the line near the number.**

_____ 1. Because fences are not _____ structures, lower knotty grades of lumber may be used to build them.

_____ 2. Parts of the fence that are set in the ground or exposed to constant moisture should be _____ or all-heart decay resistant wood.

_____ 3. When moisture comes in contact with inferior hardware or fasteners used on fences, corrosion results causing unsightly _____ on the fence.

_____ 4. Placement or height of fences sometimes is restricted by _____ regulations.

_____ 5. When pickets are applied with their edges tightly together, the assembly is called a _____ fence.

_____ 6. The board-on-board fence is similar to the picket fence except the boards are _____ from side to side.

_____ 7. The _____ fence creates a solid barrier with boards or panels fitted between top and bottom rails.

_____ 8. The _____ fence permits the flow of air through it, and yet provides privacy.

_____ 9. Because most post and rail designs have large _____, they are not intended to be used as barriers to prevent passage through them.

_____ 10. Iron fence posts should be _____ or otherwise coated to prevent corrosion.

_____ 11. The first step in building a fence is to set the _____.

_____ 12. If steep sloping land prohibits the use of a line when setting fence posts, it may be necessary to use a _____ to lay out a straight line.

_____ 13. When building on a property line, be sure the exact locations of the _____ are known.

_____ 14. Fence posts are generally set about _____ feet apart.

Name _____  Date _____

# CHAPTER 69  GYPSUM BOARD

## Multiple Choice

Write the letter for the correct answer on the line next to the number of the sentence.

_____ 1. Another name for gypsum board is _____ .
   A. drywall
   B. *Sheetrock*
   C. plasterboard
   D. A, B, and C

_____ 2. Gypsum board is composed of _____ .
   A. compressed paper coated with a gypsum surface
   B. wood fibers coated with a gypsum surface
   C. 100% gypsum
   D. a gypsum core encased in paper

_____ 3. The long edges of the most commonly used gypsum board panels are _____ .
   A. tongue and groove
   B. rabbeted
   C. tapered
   D. mitered

_____ 4. Eased edge gypsum board has a special _____ .
   A. tapered rounded edge
   B. thickened rounded edge
   C. tapered square edge
   D. thickened square edge

_____ 5. Type X gypsum board is typically known as _____ .
   A. fire code board
   B. gypsum lath
   C. red board
   D. backing board

_____ 6. Water-resistant gypsum board is easily recognized by its distinctive _____ face.
   A. brown
   B. green
   C. yellow
   D. orange

_____ 7. Blue board is the common name for _____ .
   A. gypsum lath
   B. veneer plaster base
   C. aluminum foil backed gypsum board
   D. predecorated panels

_____ 8. The most commonly used thickness of gypsum board for walls and ceilings in manufactured housing is _____.
   A. ¼"
   B. ⁵⁄₁₆"
   C. ⅜"
   D. ½"

_____ 9. _____ is used extensively as a base for ceramic tile.
   A. Coreboard
   B. Linerboard
   C. Brown board
   D. Cement board

_____ 10. Staples are an approved fastener for gypsum panels _____.
   A. in all circumstances
   B. on baselayers of multilayer applications when they penetrate the supports at least ⅝"
   C. only along tapered edges on regular gypsum board
   D. in warmer climates only

## Completion

Complete each sentence by inserting the correct answer on the line near the number.

_____ 1. The heads on gypsum board nails must be at least ____ inch in diameter.

_____ 2. A drywall hammer has a face that is _____.

_____ 3. Care should be taken to drive nails at a right angle into gypsum board panels to prevent breaking the _____.

_____ 4. Type _____ drywall screws are used for fastening into wood.

_____ 5. Type _____ drywall screws are used for fastening into gypsum backing boards.

_____ 6. Type _____ drywall screws are used for fastening into heavier gauge metal framing.

_____ 7. It is especially important to wear eye protection when driving drywall screws into _____ framing.

_____ 8. For bonding gypsum board directly to supports, special drywall _____ adhesive or approved construction adhesive is used.

_____ 9. Caution must be exercised when using some types of drywall adhesives that contain a _____ solvent.

_____ 10. When laminating gypsum boards to each other, no supplemental fasteners are needed if _____ adhesives are used.

Name _____   Date _____

# CHAPTER 70 SINGLE- AND MULTI-LAYER DRYWALL APPLICATION

## Multiple Choice

Write the letter for the correct answer on the line next to the number of the sentence.

_____ 1. Drywall should be delivered to the job site _____ .
   A. as soon as the roof is on
   B. when all the rough carpentry is complete
   C. shortly before installation begins and the building is watertight
   D. anytime after construction begins

_____ 2. Drywall should be stored _____ .
   A. on its edge leaning against a wall
   B. under cover and stacked flat on supports
   C. on three supports at least 2" wide
   D. under cover and on its edge

_____ 3. The _____ is a specially designed tool for making parallel cuts close to long edges.
   A. utility knife
   B. drywall saw
   C. drywall T-square
   D. gypsum board stripper

_____ 4. Stud edges that are to have gypsum panels installed on them must not be out of alignment more than _____ to adjacent studs.
   A. 1/16"
   B. 1/8"
   C. 1/4"
   D. 3/8"

_____ 5. When the single nailing method is used, nails are spaced a maximum of _____ on center for walls.
   A. 4"
   B. 6"
   C. 8"
   D. 10"

_____ 6. With double-nailing, the first nail must be _____ after driving the second nail of each set.
   A. removed
   B. reseated
   C. partially withdrawn
   D. set below the surface of the panel

203

_____ 7. When fastening drywall ceilings with screws to framing members that are 16" on center, the screws should be spaced _____ on center.
A. 4"
B. 8"
C. 12"
D. 16"

_____ 8. When applying adhesives to studs where two panels are joined, _____.
A. apply one straight bead to the centerline of the stud
B. zigzag the bead across the studs centerline
C. apply two parallel beads to either side of the centerline
D. cover the entire stud with adhesive

_____ 9. On ceilings where adhesive is used, the field is fastened at about _____ intervals.
A. 8"
B. 12"
C. 16"
D. 24"

_____ 10. The reason for prebowing gypsum panels is to _____.
A. compensate for misaligned studs
B. eliminate the need for adhesives
C. eliminate the need for fasteners at the top and bottom plates
D. reduce the number of supplemental fasteners required

_____ 11. _____ are supports made in the form of a "T," that are used to help hold ceiling drywall panels in position when fastening.
A. Deadmen
B. Strong backs
C. Ledger boards
D. Sleepers

_____ 12. If a fastener misses a support, _____.
A. continue driving it until it is flush with the surface of the panel
B. remove it and dimple the hole
C. set it beneath the panel's surface
D. drive its head and all clear through the panel, then patch the hole

## Completion

**Complete each sentence by inserting the correct answer on the line near the number.**

_____ 1. When walls are less than 8'-1" high, wallboard is usually installed _____.

_____ 2. The best way to minimize end joints when hanging dry wall is to use the _____ panel of drywall as possible.

_____ 3. End joints should not fall on the same _____ as those on the opposite side of the partition.

_____  4. To be less conspicuous, end joints should be as far from the _____ of the wall as possible.

_____  5. When using a drywall cutout tool, care must be taken not to plunge too deeply and contact _____ .

_____  6. When walls are more than 8'-1" high, _____ application of wallboard is more practical.

_____  7. The _____ method of drywall application helps prevent nail popping and cracking where walls and ceilings meet.

_____  8. When applying gypsum panels to curved surfaces, _____ the panels enable them to bend easier.

_____  9. _____ gypsum board and cement board panels are used in bath and shower areas as bases for the application of ceramic tile.

_____  10. With multilayer application of gypsum board, the joints of the face layer are offset at least _____ inches from the joints in the base layer.

**205**

Name _____ Date _____

# CHAPTER 71 CONCEALING FASTENERS AND JOINTS

## Multiple Choice

Write the letter for the correct answer on the line next to the number of the sentence.

_____ 1. For 24 hours before, during, and for at least 4 days after the application of joint compound, the temperature should be maintained at a minimum of _____.
   A. 40°
   B. 50°
   C. 60°
   D. 70°

_____ 2. Joints between panels that are a ¼" or more should be _____.
   A. prefilled with compound
   B. filled with insulation
   C. moistened prior to filling
   D. primed with a latex primer

_____ 3. When embedding the tape in compound, be sure _____.
   A. to center the tape on the joint
   B. there are no air bubbles under the tape
   C. there is not over 1/32" under the edges
   D. A, B, and C

_____ 4. Immediately after embedding the tape in compound, _____.
   A. moisten it with a fine spray of water
   B. wipe it with a damp sponge
   C. lift the edges and apply additional compound
   D. apply another thin coat of compound to the tape

_____ 5. The term spotting in drywall work refers to _____.
   A. the application of compound to conceal fastener heads
   B. air bubbles under the tape
   C. stains that appear as a result of using contaminated compound
   D. compound that drops to the floor during application

_____ 6. When applying compound to corner beads, _____.
   A. use the nose of the bead to serve as a guide for applying the trim
   B. apply the compound about 6" wide from the nose of the bead to a feather edge on wall
   C. each subsequent finishing coat is applied about 2" wider than the previous one
   D. A, B, and C

_____ 7. The second coat of compound is sometimes called the _____ coat.
   A. fill
   B. pack
   C. plug
   D. supply

_____ 8. Professional drywall finishers _____ .
   A. sand between each coat
   B. moisten the compound before sanding
   C. rarely sand any excess between coats
   D. only have to apply one coat

_____ 9. When finishing interior corners, _____ .
   A. a setting compound is usually applied to one side only of each corner
   B. joint reinforcing tape is not necessary
   C. the first coat is applied approximately 16" wide
   D. A, B, and C

_____ 10. For every 1,000 square feet of drywall area, ____ lbs. of conventional joint compound are needed.
   A. 105
   B. 115
   C. 135
   D. 155

## Completion

Complete each sentence by inserting the correct answer on the line near the number.

_____ 1. Joints are reinforced with _____ .

_____ 2. Exterior corners are reinforced with _____ .

_____ 3. Taping compound is used to embed and adhere tape to the board over the _____ .

_____ 4. Second and third coats over tapped joints are covered with a _____ compound.

_____ 5. An all purpose compound may be convenient, but it lacks the _____ and workability that a two-step compound system has.

_____ 6. Setting type joint compounds are only available in a _____ form.

_____ 7. Setting type joint compounds permit the _____ of drywall interiors in the same day.

_____ 8. To simplify its application to corners, joint tape has a _____ along its center.

_____ 9. Glass fiber mesh tape is available with a plain back or with an _____ backing for quick application.

_____ 10. Instead of using fasteners on corner beads, a _____ tool may be used to lock the bead to the corner.

_____ 11. Control joints are placed in large dry wall areas to relieve _____ from expansion and contraction.

Name _____   Date _____

# CHAPTER 72   TYPES OF WALL PANELING

## Completion

Complete each sentence by inserting the correct answer on the line near the number.

_____  1. The most widely used kind of sheet paneling is _____ .

_____  2. Less expensive plywood paneling is prefinished with a _____ wood grain or other design on a vinyl covering.

_____  3. The most commonly used length of paneling is the _____ foot length.

_____  4. Matching _____ is available to cover panel edges, corners, and joints.

_____  5. When exposed fastening is necessary, matching colored ring-shanked nails called _____ are used.

_____  6. _____ is a hardboard panel with a backed on plastic finish which is embossed to simulate ceramic wall tile.

_____  7. Commonly used thicknesses of hardboard paneling range from ⅛" to _____ inch(es).

_____  8. Particleboard panels must only be applied to a wall backing that is _____ .

_____  9. Unfinished particleboard paneling made from aromatic cedar chips is used to cover walls in _____ .

_____ 10. Kitchen cabinets and countertops are widely surfaced with plastic _____ .

_____ 11. _____ type laminate is used on cabinet sides and walls.

_____ 12. Regular or standard laminate is generally used on _____ surfaces.

_____ 13. Once a sheet of laminate comes in contact with the adhesive, it can no longer be _____ .

_____ 14. Most board paneling comes in a _____ inch thickness.

_____ 15. To avoid shrinkage, board paneling, like all interior finish, must be dried to a _____ content.

209

## Identification: Solid Wood Paneling Patterns

Identify each term, and write the letter of the correct answer on the line next to each number.

_____ 1. matched and eased channel

_____ 2. matched, edge and center grooved

_____ 3. matched and V-grooved

_____ 4. pickwick

_____ 5. channel rustic

_____ 6. shiplaped and V-grooved

_____ 7. tongue and grooved

_____ 8. matched, V-grooved and beaded

Name _____ Date _____

# CHAPTER 73  APPLICATION OF WALL PANELING

## Multiple Choice

Write the letter for the correct answer on the line next to the number of the sentence.

_____ 1. Sheet paneling is usually applied to walls with the long edges _____ .
   A. vertical
   B. horizontal
   C. diagonal
   D. tight against the ceiling

_____ 2. The installation of a gypsum board base layer beneath sheet paneling is done so that _____ .
   A. the wall is stronger and more fire resistant
   B. sound transmission is deadened
   C. there will be a rigid finished surface for application of paneling
   D. A, B, and C

_____ 3. Before the installation of paneling to masonry walls, _____ .
   A. contact adhesive must be applied to the masonry
   B. furring strips must be applied to the masonry
   C. the wall must be parged with portland cement
   D. the back of the paneling sheet must be sealed

_____ 4. Paneling edges must fall _____ .
   A. midway between studs
   B. in line with the left side of the stud
   C. in line with the right side of the stud
   D. on stud centers

_____ 5. Panels are usually fastened with _____ .
   A. screws
   B. finish nails
   C. contact adhesive
   D. color pins and adhesive

_____ 6. Paneling that only covers the lower portion of the wall is called _____ .
   A. wainscoting
   B. louvering
   C. ledgering
   D. garthcoting

_____ 7. When cutting to a scribed line that must be made on the face side of prefinished paneling, it is recommended that a _____ be used.
   A. circular saw
   B. hand ripsaw
   C. fine-toothed hand crosscut saw
   D. saber saw

211

_____ 8. When using adhesive on paneling, _____.
   A. apply beads about 4" long and 16" apart where panel edges and ends make contact
   B. do not allow the panel to move once it has made contact with the wall
   C. apply a ⅛" continuous bead to the intermediate studs only
   D. after the initial contact is made between the wall and the panel, the sheet is pulled a short distance away from the wall and then pressed back into position

_____ 9. When cutting openings for wall outlet, _____.
   A. a saber saw may be used if the cut is made from the back of the panel
   B. a circular saw may be used
   C. the opening may be cut oversized to allow for error
   D. a ripsaw is used

_____ 10. When applying paneling to exterior corners, _____ corner molding may be used.
   A. wood
   B. metal
   C. vinyl
   D. A, B, and C

## Completion

Complete each sentence by inserting the correct answer on the line near the number.

_____ 1. For vertical application of board paneling to a frame wall, _____ must be provided between the studs.

_____ 2. Prior to its application, board paneling should stand against the walls around the room to allow it to adjust to room temperature and _____.

_____ 3. If tongue and groove board paneling is used, tack the first board in a plumb position with the _____ edge in the corner.

_____ 4. When fastening tongue and groove board paneling, blind nail into the _____ only.

_____ 5. If board paneling is of uniform width, the _____ of the first board must be planned to avoid ending with a small strip.

_____ 6. When fastening the last piece of vertical board siding, the cut edge goes in the _____.

_____ 7. Blocking between studs on open walls is not necessary when siding is applied _____.

_____ 8. Specially matched _____ is used between panels and on interior and exterior corners of plastic laminates that is prefabricated to plywood sheets.

## Discussion

Write your answer on the lines below the instruction.

1. Describe the process of estimating the number of sheets of paneling needed to do a room.

_____

_____

_____

Name _____     Date _____

# CHAPTER 74 CERAMIC WALL TILE

## Multiple Choice

Write the letter for the correct answer on the line next to the number of the sentence.

_____ 1. The proper backing for ceramic wall tile is _____ .
   A. water-resistant gypsum or cement board
   B. hard board
   C. plywood
   D. A, B, and C

_____ 2. The most commonly used wall tiles are nominal size _____ squares, in about ¼" thickness.
   A. 1" and 3"
   B. 2" and 5"
   C. 3½" and 4½"
   D. 4" and 6"

_____ 3. When tile is installed around a tub that lacks a showerhead, it should extend a minimum of _____ above the rim.
   A. 4"
   B. 6"
   C. 8"
   D. 10"

_____ 4. Around tubs with showerheads, the tile should extend a minimum of _____ above the rim or 6" above the showerhead, whichever is higher.
   A. 3'
   B. 4'
   C. 5'
   D. 6'

_____ 5. Before beginning the application of ceramic wall tile, the width of the _____ must be determined.
   A. border tile
   B. field tile
   C. application trowel
   D. tile saw kerf

_____ 6. When troweling adhesive to the wall in preparation for tile application, _____ .
   A. it is best to go on the heavy side
   B. a regular straight edge trowel is to be used
   C. overlap the coverage area with the adhesive
   D. be sure the trowel is the one recommended by the manufacturer

213

_____ 7. Whole tiles that are applied to the center of the wall are called _____ tiles.
   A. bullnose
   B. counter
   C. field
   D. court

_____ 8. A hand-operated ceramic tile cutter works in a manner that is similar to a _____.
   A. glass cutter
   B. masonry saw
   C. rasp
   D. coping saw

_____ 9. After all tile has been applied, the joints are filled with _____.
   A. portland cement
   B. tile grout
   C. tile mastic
   D. contact cement

_____ 10. After the joints are partially set up but not completely hardened, they then must be _____.
   A. pointed
   B. checked
   C. glazed
   D. pitched

Name _____  Date _____

# CHAPTER 75  SUSPENDED CEILINGS

## Multiple Choice

Write the letter for the correct answer on the line next to the number of the sentence.

_____ 1. A main advantage of using a suspended ceiling is that the space above it can be utilized for _____ .
   A. recessed lighting
   B. duct work
   C. pipes and conduit
   D. A, B, and C

_____ 2. Suspended ceilings systems consist of panels that are _____ .
   A. stapled into firing strips
   B. applied with adhesive
   C. laid into a metal grid
   D. tongue and groove that interlock into each other

_____ 3. L-shaped pieces that are fastened to the wall to support the ends of main runners and cross tees are called _____ .
   A. wall ties
   B. wall angles
   C. support angles
   D. runner supports

_____ 4. Main runners are shaped in the form of _____ .
   A. an L
   B. a T
   C. an upside down L
   D. an upside down T

_____ 5. Slots are punched in the side of the runners at _____ intervals to receive cross tees.
   A. 4"
   B. 8"
   C. 12"
   D. 16"

_____ 6. The primary support for the ceiling's weight are the _____ .
   A. main runners
   B. cross tees
   C. ceiling panels
   D. wall ties

_____ 7. Panels come in 2' × 2' and 2' × 4' sizes with square and _____ edges.
   A. rounded
   B. beveled
   C. concaved
   D. rabbeted

215

_____ 8. Main runners are usually spaced _____ apart.
   A. 2'
   B. 4'
   C. 6'
   D. 8'

_____ 9. The first part of the ceiling grid system to be installed are the _____.
   A. main runners
   B. cross tees
   C. wall angles
   D. wall ties

_____ 10. A suspended ceiling must be installed with at least _____ clearance below the lowest air duct, pipe, or beam for enough room to insert ceiling panels in the grid.
   A. 1"
   B. 3"
   C. 5"
   D. 8"

## Completion

Complete each sentence by inserting the correct answer on the line near the number.

_____ 1. Screw eyes are to be installed not over ____ feet apart.

_____ 2. Screw eyes must be long enough to penetrate wood joists by at least ___ inch.

_____ 3. For residential work _____ gauge hanger wire is usually used.

_____ 4. About 6" of hanger wire is inserted through the screw eye and then securely wrapped around itself _____ times.

_____ 5. A cross tee line must be stretched across the short dimension of the room to line up the _____ in the main runners.

_____ 6. Cross tees are installed by inserting the tabs on the ends into the slots in the _____.

_____ 7. Install all full-size field panels first and then cut and install _____ panels.

_____ 8. Panels are cut with a sharp _____.

_____ 9. Always cut ceiling panels with their finish side placed _____.

_____ 10. To find the number of wall angle needed in a room, divide the perimeter of the room by _____.

Name _____  Date _____

# CHAPTER 76 CEILING TILE

## Completion

Complete each sentence by inserting the correct answer on the line near the number.

_____  1. _____ fiber tiles are the lowest in cost.

_____  2. _____ fiber tiles are used when a more fire-resistant tile is required.

_____  3. The most popular size square tile is the _____ inch.

_____  4. The tiles that run along the walls are known as _____ tiles.

_____  5. If the short wall of a room measures 12'-6", then the border tiles along the long walls of the room would measure _____ .

_____  6. When tiles are not being applied with adhesive to an existing ceiling, then _____ must be installed to fasten the tiles to.

_____  7. Ceiling joists are usually _____ inches on center.

_____  8. Ceiling joists usually run parallel to the _____ dimension of the room.

_____  9. If the size of the tiles are 12", then the furring strips must be installed _____ inches on center.

_____  10. Furring strip fasteners must penetrate at least _____ inch(es) into the joist.

_____  11. Prior to installing, ceiling tiles should be allowed to adjust to normal interior room conditions for _____ hours.

_____  12. To help prevent fingerprints and smudges on the finished ceiling, some carpenters sprinkle _____ on their hands.

_____  13. On border tiles, all _____ edges should go against the wall.

_____  14. The correct number of ½" or 9/16" staples to use in 12" square ceiling tiles is _____ .

_____  15. With 12" × 24" tiles, the correct number of staple to use in each tile is ___ .

_____  16. When applying adhesive to ceiling tile, each daub of it should be about the size of a _____ .

_____  17. When using a concealed grid system, the tiles edges must be _____ .

_____  18. With the concealed grid system, the tiles in the end row are held tight by inserting a _____ between each tile and the end row.

217

# Math

**Estimate the number of 12″ × 24″ ceiling tile and the lineal feet of furring strips needed to cover the ceiling of a room that measures 12′-8″ by 23′-4″.**

Name _____  Date _____

# CHAPTER 77  DESCRIPTION OF INTERIOR DOORS

## Completion

**Complete each sentence by inserting the correct answer on the line near the number.**

_____  1. The interior _____ door is used when a less expensive smooth surfaced door with a plain appearance is desired.

_____  2. Most interior residential doors are manufactured in a _____ inch thickness.

_____  3. The most common height of manufactured doors is _____.

_____  4. Door widths range from 1'-0" to _____ in increments of two inches.

_____  5. Solid core doors are generally used on _____ doors.

_____  6. The _____ is the thin plywood that covers the frame and mesh of a hollow core door.

_____  7. Doors that swing in both directions are known as _____ doors.

_____  8. _____ doors are not practical in openings less than 6' wide.

_____  9. _____ doors require more time and material to install than other methods of door operation.

_____  10. Bifold doors on the jamb side swing on _____ , installed at the top and bottom.

## Matching

**Write the letter for the correct answer on the line near the number to which it corresponds.**

_____  1. lauan plywood          A. obstruct vision but permit the flow of air

_____  2. folding doors          B. rides on rollers in a double track

_____  3. french doors           C. used extensively for flush door skins

_____  4. louver doors           D. may contain from 1 to 15 lights of glass

_____  5. bypass door            E. hung in pairs that swing in both directions

_____  6. pocket doors           F. only the door's lock edge is visible when opened

_____  7. cafe doors             G. consists of many narrow panels each about the same width as the door jamb

## Discussion

**Write your answer on the lines below the instruction.**

1. What are some of the special conditions present in heavy commercial buildings that would require heavier and larger interior doors than those used in homes?

_____

_____

_____

Name _____ Date _____

# CHAPTER 78  INSTALLATION OF INTERIOR DOORS AND DOOR FRAMES

## Multiple Choice

Write the letter for the correct answer on the line next to the number of the sentence.

_____ 1. The first step in making an interior door frame is to _____.
   A. install the stop
   B. check the width and height of the rough opening
   C. rabbet the edge of the jamb stock
   D. install the header jamb

_____ 2. The rough opening width for single acting swinging doors should be the door width plus _____.
   A. the thickness of the jamb plus 1"
   B. double the door thickness
   C. double the side jamb thickness plus ½"
   D. triple the side jamb thickness plus ¾"

_____ 3. The rough opening height of the door should be the door's height plus _____, plus the thickness of the finish floor, plus the desired clearance under the door.
   A. the thickness of the header jamb plus ¼"
   B. double the header jamb thickness
   C. double the side jamb thickness plus ½"
   D. ½"

_____ 4. Interior door frames are usually installed _____.
   A. during the rough framing process
   B. after the interior wall covering is applied
   C. at the same time the interior wall covering is applied
   D. prior to the application of the interior wall covering

_____ 5. To find the width of the jamb stock, _____.
   A. triple the thickness of the door
   B. subtract the thickness of the door jamb from that of the total wall thickness
   C. measure the total wall thickness including the wall covering
   D. subtract 1" from the sill width

_____ 6. Door frames must be set so that the jambs are _____.
   A. straight
   B. level
   C. plumb
   D. A, B, and C

_____ 7. If a rabbeted frame is used, _____.
   A. the door's swing must be determined so that the rabbet faces the right direction
   B. a separate stop needs to be applied to the inside of the door frame
   C. the rough opening is the same width and height as the door frame
   D. the horns are not to be cut from the top edge of the side jambs

_____ 8. The header jamb is leveled by placing shims between _____.
   A. the header jamb and the header
   B. the header and the horns
   C. the bottom of the appropriate side jamb and the subfloor
   D. side jamb opposite the header jamb

_____ 9. When straightening side jambs by shimmying at intermediate points use a _____.
   A. combination square
   B. 6' straight edge
   C. butt gauge
   D. sliding T-bevel

_____ 10. Before any nails are driven home when setting a door frame, the frame should be checked for a _____.
   A. wind
   B. bluster
   C. gale
   D. breeze

## Completion

Complete each sentence by inserting the correct answer on the line near the number.

_____ 1. Door stops are not permanently fastened until any necessary adjustment is made when the _____ is installed.

_____ 2. The special pivoting hardware installed on double acting doors returns the door to a _____ position after being opened.

_____ 3. Bypass door tracks are installed on the _____ according to the manufacturer's directions.

_____ 4. It is important that the door pulls on bypass doors are installed _____ so as not to obstruct the bypassing door.

_____ 5. Before installing bifold door tracks into position, be sure the _____ for the door pivot pins are inserted in the track.

_____ 6. Pocket door frames are usually assembled at the _____.

_____ 7. Because folding doors are made by many different manufacturers and come in many different styles, it is important to closely follow the _____ supplied with the unit.

_____ 8. To accommodate various wall thicknesses, prehung door units are available in various _____ widths.

_____ 9. To maintain proper clearance between the door and the frame on prehung small cardboard _____ are stapled to the lock edge and the top end of the door.

_____ 10. The _____ lock is used often on bathroom and bedroom doors.

Name _____    Date _____

# CHAPTER 79   DESCRIPTION AND APPLICATION OF MOLDING

## Completion

Complete each sentence by inserting the correct answer on the line near the number.

_____ 1. In order to present a suitable appearance, moldings must be applied with _____ joints.

_____ 2. To reduce waste, door casings are available in lengths of _____ feet.

_____ 3. Finger-jointed lengths of molding should only be used when a _____ finish is to be applied.

_____ 4. Moldings are either classified by their _____ or their designated location.

_____ 5. The joint between the bottom of the base and the finish floor is usually concealed by the base _____.

_____ 6. _____ are used to trim around windows, doors, and other openings to cover the space between the wall and the frame.

_____ 7. For a more decorative appearance, _____ are applied to the outside edges of casings.

_____ 8. Aprons and stools are part of _____ trim.

_____ 9. Another name for outside corners are corner _____.

_____ 10. Caps and chair rails are used to trim the top edge of _____.

_____ 11. With the exception of prefinish molding, joints between molding lengths should be _____ flush after the molding has been fastened.

_____ 12. Molding joints on exterior corners must be _____.

_____ 13. Joints on interior corners, especially on large moldings, are usually _____.

_____ 14. Another name for the power miter box is the _____ saw.

_____ 15. When cutting molding in a miter box, the face side of the molding should be placed _____.

_____ 16. When mitering bed, crown, and cove moldings, place them _____ in the miter box.

_____ 17. The _____ is the tool to use when paper-thin corrective cuts are needed when fitting miter joints.

223

_____ 18. A mitering _____ may be used to make miters on a table or radial arm saw quickly and easily without any change in the setup.

_____ 19. When making a coped joint, the coping saws handle should be above the work, and its teeth should face _____ from the saw's handle.

_____ 20. To assure straight application of large size ceiling moldings, a _____ should be used as a guide.

_____ 21. Nails should be placed 2" to 3" from the molding's end to prevent _____.

## Identification: Molding

Identify each term, and write the letter of the correct answer on the line next to each number.

_____ 1. crown

_____ 2. bed

_____ 3. cove

_____ 4. quarter round

_____ 5. corner guard

_____ 6. base shoes

_____ 7. chair rail

_____ 8. base moldings

_____ 9. half round

_____ 10. hand rail

Name _____     Date _____

# CHAPTER 80  APPLICATION OF DOOR CASINGS, BASE, AND WINDOW TRIM

## Multiple Choice

Write the letter for the correct answer on the line next to the number of the sentence.

_____ 1. Door casing must be applied before _____.
   A. window trim
   B. ceiling molding
   C. base moldings
   D. corner guards

_____ 2. _____ blocks are small decorative blocks used as part of the door trim at the base and at the head.
   A. Plinth
   B. Batten
   C. Astragal
   D. Crown

_____ 3. Molded casings usually have their back sides _____.
   A. rabbeted
   B. backed out
   C. beveled
   D. laminated

_____ 4. An alternative to using molded casings is to use _____ stock.
   A. S4S
   B. S3S
   C. S2S
   D. S1S

_____ 5. The 3/16" to 5/16" setback of door casings from the inside face of the door frame is called the _____.
   A. reveal
   B. exposure
   C. rake
   D. salvage

_____ 6. For each interior door opening, _____ are required.
   A. two side casings and two header casings
   B. two side casings and four header casings
   C. four side casings and four header casings
   D. four side casings and two header casings

_____ 7. When fastening casing into _____, use 6d or 8d finish nails.
   A. header jambs
   B. side jambs
   C. framing
   D. A, B, and C

**225**

_____ 8. To bring the faces of mitered casing joints flush _____.
   A. sand them
   B. shim between the casing back and the wall
   C. use a wood chisel to remove thin shavings from the thicker side of the joint
   D. use a flat wood file

_____ 9. Base trim should be _____.
   A. thinner than the door casing
   B. molded or S2S stock
   C. back mitered to outline the cope for interior corners
   D. A, B, and C

_____ 10. When fastening base molding, use _____ finishing nail(s) of sufficient length at each stud location.
   A. one
   B. two
   C. three
   D. four

## Completion

**Complete each sentence by inserting the correct answer on the line near the number.**

_____ 1. When scribing lines on baseboard, be sure to hold the dividers so that a line between the two points is _____ to the floor.

_____ 2. When cutting baseboard to a scribed line, be sure to give the cut a slight _____.

_____ 3. On outside corners of baseboard, make regular _____ joints.

_____ 4. The base shoe is usually nailed into the _____.

_____ 5. No base shoe is required if _____ is used as a finish floor.

_____ 6. The bottom side of the stool is _____ at an angle, so its top side will be level after it is fit on the window's sill.

_____ 7. The _____ covers the joint between the sill and the wall.

_____ 8. Jamb _____ must be used when windows are installed with jambs that are narrower than the walls.

_____ 9. Window casings are installed with their inside edges flush with the inside face of the _____.

_____ 10. The _____ is a piece of 1" × 5" stock installed around the walls of a closet to support the rod and the shelf.

_____ 11. Closet pole sockets should be located at least _____ inches from the back wall.

_____ 12. When sanding interior trim, always sand with the _____.

_____ 13. All traces of excess glue must be removed if the trim is to be _____.

_____ 14. To help prevent the hammer from glancing off the head of a nail, occasionally clean the hammer's face by rubbing it with _____.

Name _____   Date _____

# CHAPTER 81  DESCRIPTION OF STAIR FINISH

## Multiple Choice

Write the letter for the correct answer on the line next to the number of the sentence.

_____ 1. The _____ is a component of the stair body finish.
   A. newel post
   B. riser
   C. handrail
   D. baluster

_____ 2. With an open staircase, _____ .
   A. tread ends butt against the wall
   B. one or more ends of the treads are exposed to view
   C. the area under the stair body is exposed to view
   D. both sides of the treads must be exposed to view

_____ 3. The part of the stair tread that extends beyond the riser is the _____ .
   A. nosing
   B. tread molding
   C. stringer
   D. half newel

_____ 4. Finish stringers are sometimes called _____ .
   A. preachers
   B. skirt boards
   C. carriages
   D. aprons

_____ 5. Open finish stringers are placed _____ .
   A. on the open side of the stairway above the treads
   B. on the closed side of an open stairway above the treads
   C. on the open side of a stairway below the treads
   D. on the closed side of an open stairway below the treads

_____ 6. When the staircase is open on one or more sides _____ .
   A. a starting step may be used
   B. return nosing is used
   C. an open finish stringer is used
   D. A, B, and C

_____ 7. In post to post balustrades _____ .
   A. the newel post fits into the bottom of the handrail
   B. the newel posts have flat square surfaces near the top
   C. goosenecks must always be used
   D. the newel posts are made with a pin in their top

**227**

_____ 8. Narrow strips called _____ are used between balusters to fill the plowed groove on handrails and shoe rails.
   A. fillet
   B. rossettes
   C. mullions
   D. scabs

_____ 9. Short sections of specialty curved handrail are called _____ .
   A. stiles
   B. fittings
   C. heels
   D. crickets

_____ 10. _____ are vertical, usually decorative pieces between newel posts and spaced close together supporting the handrail.
   A. Voltues
   B. Cleats
   C. Battens
   D. Balusters

# Identification: Balustrade Components

Identify each term, and write the letter of the correct answer on the line next to each number.

_____ 1. baluster

_____ 2. rake handrail

_____ 3. landing newel

_____ 4. closed finish stringer

_____ 5. half newel

_____ 6. starting newel

_____ 7. landing rail

_____ 8. balcony handrail

_____ 9. landing baluster

_____ 10. balcony baluster

Name _____   Date _____

# CHAPTER 82  FINISHING THE STAIR BODY OF OPEN AND CLOSED STAIRCASES

## Multiple Choice

Write the letter for the correct answer on the line next to the number of the sentence.

_____ 1. Use a _____ when cutting the level and plumb cuts on a closed finish stringer.
   A. 7- or 8-point crosscut handsaw
   B. fine tooth crosscut handsaw
   C. 5½-point hand ripsaw
   D. coping saw

_____ 2. The nosed edge of the tread extends beyond the face of the riser by _____.
   A. ⅝"
   B. ⅞"
   C. 1⅛"
   D. 1⅝"

_____ 3. Prior to placing the treads between the finish stringers _____.
   A. rub wax on one end
   B. liberally apply glue to the tread's end grain
   C. spray the treads with a light mist of water
   D. seal the treads with a varnish

_____ 4. When fastening treads, to prevent splitting the wood _____.
   A. predrill the nail holes
   B. use only 12d finish nails
   C. be sure to use galvanized 10d casing nails
   D. use only glue and no nails

_____ 5. Tread molding is attached using _____ finish nails.
   A. 3d
   B. 4d
   C. 6d
   D. 8d

_____ 6. On a staircase that has one side open and the other side closed, the _____ the first item of finish to be applied.
   A. risers are
   B. closed finish stringer is
   C. open finish stringer is
   D. treads are

_____ 7. When laying out the plumb lines on an open finish stringer, a helpful device to use is a _____.
   A. chalk line
   B. plumb bob
   C. line level
   D. preacher

_____ 8. When fastening a mitered riser to a mitered stringer, _____.
   A. sand all pieces before installation
   B. apply a small amount of glue to the joint
   C. drive finish nails both ways through the miter
   D. A, B, and C

_____ 9. When ripping treads to width, _____.
   A. bevel the back edge
   B. make allowances for the rabbeted back edge
   C. bevel the front edge of the tread
   D. only hand ripsaw should be used

_____ 10. Treads on winding steps _____.
   A. are especially difficult to fit
   B. have the same angle on both ends
   C. are wider on the inside than the out
   D. A, B, and C

## Completion

Complete each sentence by inserting the correct answer on the line near the number.

_____ 1. Instead of a framing square, a _____ is often used laying out a housed stringer.

_____ 2. The _____ is joined to the housed stringer at the top and bottom of the staircase.

_____ 3. A _____ is used to guide the router when housing the stringer.

_____ 4. Housed stringers are routed so that the dadoes are the exact width at the nosing and wider toward the inside so the treads and risers can be _____ against the shoulders of the dadoes.

_____ 5. An opened stringer is also referred to as a _____ stringer.

_____ 6. The vertical layout line on an opened stringer is mitered to fit the mitered end of the _____ .

_____ 7. On open stringers, a _____ cut is made through the stringer's thickness for the tread.

_____ 8. A ⅜" by ⅜" groove is cut on the face side of all but the first riser, to receive the rabbeted inner edge of the _____ .

_____ 9. If the staircase is open, a return _____ is mitered to the end of the tread.

_____ 10. _____ are usually the first members applied to the stair carriage in a closed staircase.

_____ 11. The top edge of a closed finished stringer is usually about _____ inches above the tread nosing.

_____ 12. When cutting along the layout lines for a closed finished stringer, extreme care must be made on the _____ cut.

Name _____    Date _____

# CHAPTER 83  BALUSTRADE INSTALLATION

## Multiple Choice

**Write the letter for the correct answer on the line next to the number of the sentence.**

_____ 1. The first step in laying out balustrades is to _____.
   A. lay out the handrail
   B. determine the rake of the handrail
   C. lay out the balustrade's centerline
   D. determine the height of the starting newel

_____ 2. Most building codes require that balusters be spaced so that _____.
   A. a small child's foot cannot fit between them
   B. there is no more than one per tread
   C. the space between them equals their width
   D. no object 5" in diameter can pass through

_____ 3. On open treads, the center of the front baluster is located _____.
   A. a distance equal to ½ its thickness, back from the riser's face
   B. flush with the riser's face
   C. on the center of the tread's outside edge

_____ 4. Rake handrail height is the vertical distance from the tread nosing _____.
   A. to the handrail bottom side
   B. to the handrail center
   C. to the top of the handrail
   D. to the top of the newel post

_____ 5. On a stairway more than 88" in width, _____.
   A. handrail is needed on only one side
   B. a handrail is needed on both sides
   C. a handrail must be provided in the stairway's center
   D. B and C

_____ 6. The starting newel post is notched over _____.
   A. the inside corner of the second step
   B. the inside corner of the first step
   C. the outside corner of the first step
   D. the outside corner of the second step

_____ 7. Generally, codes require that balcony rails for homes be not less than _____.
   A. 18"
   B. 24"
   C. 36"
   D. 42"

233

_____ 8. To determine the height of a balcony newel, you must know _____.
   A. the height of the balcony handrail plus 1" for a block reveal
   B. the height of the newels turned top
   C. the distance the newel extends below the floor
   D. A, B, and C

_____ 9. When square top balusters are used, _____.
   A. holes must be bored in the bottom side of the handrail at least ¾" deep
   B. the balusters must be trimmed to length at the rake angle
   C. holes need not be bored in the treads
   D. A, B, and C

_____ 10. Instead of a half newel, a _____ is sometimes used to end the balcony handrail.
   A. landing fitting
   B. gooseneck
   C. half baluster
   D. rosette

## Completion

**Complete each sentence by inserting the correct answer on the line near the number.**

_____ 1. Square top balusters are fastened to the handrail with finish nails and _____ .

_____ 2. When using square top balusters, _____ are installed between the balusters in the plow of the handrail.

_____ 3. When installing an over-the-post balustrade, the first step is to lay out the balustrade and baluster _____ on the stair treads.

_____ 4. A _____ is a piece of wood cut in the shape of a right triangle, whose sides are equal in length to the rise and tread run of the stairs.

_____ 5. When laying out the starting fitting, mark it at the _____ point, where its curve touches the pitch block.

_____ 6. When cutting handrail fittings on a power miter box, be sure to securely _____ the fitting and the pitch block.

_____ 7. To mark the hole locations for handrail bolts, a _____ should be used to assure proper alignment.

_____ 8. A 1-riser balcony gooseneck fitting is used when the balcony rails are _____ inches high.

_____ 9. On over-the-post balustrades, the height of the rake handrail is calculated from the height of the starting _____ .

_____ 10. The height of a balcony newel on an over-the-post balustrade is found by subtracting the handrail thickness from the handrail _____ .

Name _____  Date _____

# CHAPTER 84  DESCRIPTION OF WOOD FINISH FLOORS

## Multiple Choice

**Write the letter for the correct answer on the line next to the number of the sentence.**

_____ 1. Most hardwood finish flooring is made from _____.
   A. Douglas fir
   B. white or red oak
   C. hemlock
   D. southern yellow pine

_____ 2. The most widely used type of solid wood flooring is _____.
   A. strip
   B. laminated parquet blocks
   C. laminated strip
   D. parquet strip

_____ 3. Unfinished strip flooring _____.
   A. has a chamfer machined between the face and edge sides
   B. cannot be sanded after installation
   C. is milled with square sharp corners at the intersection of the face and the edges
   D. are waxed at the factory

_____ 4. Laminated strip flooring _____.
   A. is easily recognizable by the V-grooves on the floor's surface after it is laid
   B. is a 5-ply prefinished wood assembly
   C. must be sanded after it is installed
   D. is chamfered on its edges

_____ 5. Plank flooring is similar to _____ flooring.
   A. strip
   B. parquet strip
   C. Monticello
   D. parquet strip

_____ 6. The highest quality parquet block flooring is made with _____ thick, tongue and grooved solid hardwood flooring.
   A. ⅜"
   B. ½"
   C. ⅝"
   D. ¾"

_____ 7. Monticello is the name of a parquet originally designed by _____.
   A. Benjamin Franklin
   B. George Washington
   C. Thomas Jefferson
   D. Samuel Adams

_____ 8. Laminated blocks are generally made of 3-ply laminated oak in a _____ thickness.
   A. ⅜"
   B. ½"
   C. ⅝"
   D. ¾"

_____ 9. Finger blocks are another name for _____.
   A. laminated blocks
   B. slat blocks
   C. unit blocks
   D. A, B, and C

_____ 10. _____ is the top grade of unfinished oak flooring.
   A. No. 1
   B. No. 2
   C. Clear
   D. Select

## Completion

Complete each sentence by inserting the correct answer on the line near the number.

_____ 1. In addition to appearance, grades are based on _____.

_____ 2. Red grades of unfinished pecan flooring contain all _____.

_____ 3. White grades of unfinished pecan flooring contain all _____.

_____ 4. The lowest grade of unfinished hard maple flooring is called _____ grade.

_____ 5. The average length of clear bundles of flooring is _____ feet.

_____ 6. A bundle of flooring may contain pieces from 6" under, to 6" over the _____ length of the bundle.

_____ 7. No pieces shorter than _____ inches are allowed in a bundle.

_____ 8. _____ is the lowest grade of prefinished flooring.

Name _____   Date _____

# CHAPTER 85  LAYING WOOD FINISH FLOORS

## Multiple Choice

**Write the letter for the correct answer on the line next to the number of the sentence.**

_____ 1. When fastening flooring to ½" plywood subfloors, it is required that _____ .
   A. screws be used as fasteners
   B. the fasteners penetrate into the joists
   C. rigid insulation be placed over the subfloor
   D. A, B, and C

_____ 2. When laying the first course of strip flooring, it is placed _____ from the starting wall.
   A. with its tongue side ¾"
   B. with its groove side ½"
   C. with its tongue side ½"
   D. with its groove side ¾"

_____ 3. When blind nailing flooring, drive the nails at about a _____ angle.
   A. 90°
   B. 60°
   C. 45°
   D. 30°

_____ 4. Floor layers usually use _____ to set hardened flooring nails.
   A. the head of the next nail to be driven
   B. a nail set
   C. a screwdriver laid on its side
   D. a center punch

_____ 5. Laying out loose flooring ahead of time to assure efficient installation is known as _____ .
   A. cribbing the stock
   B. racking the floor
   C. grouping the strips
   D. clustering the stock

_____ 6. When using a power nailer to fasten flooring, _____ .
   A. an air compressor must be used
   B. holes for the fasteners must be predrilled
   C. one blow must be used to drive the fastener
   D. the floor layer must not stand on the flooring

_____ 7. The last course of strip flooring must be _____ .
   A. fastened with a power nailer
   B. must be blind nailed
   C. face nailed
   D. glued and not nailed

_____ 8. Laminated strip flooring _____.
   A. needs an ⅛" foam underlayment under it
   B. is not fastened or cemented to the floor
   C. is brought up tight against the previous course with a hammer and tapping block
   D. A, B, and C

_____ 9. When installing parquet flooring, it is usually _____.
   A. face nailed
   B. blind nailed
   C. laid in mastic
   D. fastened with a power nailer

_____ 10. When laying unit blocks in a square pattern, two layout lines are snapped _____.
   A. at right angles to each other and diagonal to the walls
   B. with one parallel and one diagonal to the walls
   C. at right angles to each other parallel to the walls
   D. parallel to each other and diagonal to the walls

## Completion

**Complete each sentence by inserting the correct answer on the line near the number.**

_____ 1. The installation of wood finish floors on _____ slabs is not recommended.

_____ 2. New concrete slabs should be allowed to age at least _____ days prior to the installation of a wood finish floor.

_____ 3. When preforming a rubber mat moisture test on a concrete slab, allow the mat to remain in place at least _____ hours.

_____ 4. To ensure a trouble-free finish floor installation, a _____ must be installed over all concrete slabs.

_____ 5. Prior to the application of polyethylene film to a concrete slab, a skim coat of _____ is troweled over the entire area.

_____ 6. Exterior grade sheathing plywood may be used for a subfloor, if it is at least ____ inch thick.

_____ 7. Strips of wood laid over a concrete floor that finish flooring is attached to, are known as _____.

_____ 8. The National Oak Flooring Manufacturers Association does not recommend fastening finish flooring to subfloors of _____ panels.

_____ 9. For its best appearance, strip flooring should be laid in the direction of the room's _____ dimension.

_____ 10. To help keep out dust, prevent squeaks, and retard moisture from below, _____ is applied over the subflooring.

## Discussion

**Write your answer on the lines below the instruction.**

1. Describe the recommended procedures that are necessary to maintain the proper moisture content of hardwood flooring prior to its installation.

   _____

   _____

   _____

Name _____   Date _____

# CHAPTER 86 UNDERLAYMENT AND RESILIENT TILE

## Completion

Complete each sentence by inserting the correct answer on the line near the number.

_____   1. _____ is installed on top of the subfloor to provide a base for the application of resilient sheet or tile flooring.

_____   2. All joints between the subfloor and the underlayment should be _____ .

_____   3. To allow for expansion, leave about _____ inch between underlayment panels.

_____   4. Unless underlayment is installed over a board subfloor, its face grain should run _____ the floor joists.

_____   5. If _____ is used as a subfloor no underlayment is needed.

_____   6. Resilient floor tiles are applied to the floor in a manner similar to applying _____ .

_____   7. Long strips called _____ strips are used between the tiles to create unique floor patterns.

_____   8. Caution must be exercised when removing existing resilient floor covering because it may contain _____ .

_____   9. Before installing resilient tile, make sure underlayment _____ are not projecting above the surface.

_____   10. If border tiles are a half width or more, then the layout lines are _____ .

_____   11. Peel-and-stick floor tiles are manufactured with the _____ applied at the factory.

_____   12. When applying adhesive for floor tile, it is important that the trowel has the proper size _____ .

_____   13. When laying tiles, start at the _____ of the layout lines.

_____   14. Tiles are to be laid in place instead of _____ them into position.

_____   15. Border tiles may be cut by scoring with a sharp _____ and bending.

_____   16. Many times a vinyl _____ is used to trim a tile floor.

_____   17. The number of 12" by 12" tiles needed to cover a floor is equal to the floor's _____ in square feet.

241

## Discussion

**Write your answer on the lines below the instruction.**

1. Unless we know for sure that it does not, why should we assume that existing flooring contains asbestos?

   _____

   _____

   _____

Name _____  Date _____

# CHAPTER 87 DESCRIPTION AND INSTALLATION OF MANUFACTURED CABINETS

## Multiple Choice

Write the letter for the correct answer on the line next to the number of the sentence.

_____ 1. The method of cabinet construction that has a traditional look is called _____.
   A. face-framed
   B. European
   C. modular
   D. prosaic

_____ 2. The two basic kinds of kitchen cabinet units are _____.
   A. supporting and shelf
   B. standing and attached
   C. base and wall
   D. case and drawer

_____ 3. Countertops are usually _____ from the floor.
   A. 28"
   B. 32"
   C. 36"
   D. 40"

_____ 4. The usual overall height of a kitchen cabinet installation is _____.
   A. 6'-6"
   B. 7'-0"
   C. 7'-6"
   D. 8'-0"

_____ 5. Standard wall cabinets are _____ deep.
   A. 8"
   B. 10"
   C. 12"
   D. 14"

_____ 6. The usual countertop thickness is _____.
   A. ¾"
   B. 1"
   C. 1¼"
   D. 1½"

_____ 7. A recess called a _____ is provided at the bottom of the base cabinet.
   A. toe kick
   B. foot area
   C. counter base
   D. floor cove

_____ 8. Cabinets that provide access from both sides are called _____.
   A. dual-sided
   B. twin-entry
   C. bi-frontal
   D. double-faced

_____ 9. Double door pantry cabinets are made _____ wide.
   A. 28"
   B. 36"
   C. 42"
   D. 48"

_____ 10. Wall cabinets with a 24" depth are usually installed _____.
   A. when a contemporary appearance is desired
   B. above refrigerators and tall cabinets
   C. above ranges
   D. A, B, and C

## Completion

Complete each sentence by inserting the correct answer on the line near the number.

_____ 1. Most vanity base cabinets are manufactured _____ inches high.

_____ 2. The first step in drawing a cabinet layout plan is to carefully and accurately _____ the walls on which the cabinets are to be installed.

_____ 3. Many large kitchen cabinet distributors will, on request, provide _____.

_____ 4. Scribbing and fitting cabinets to an uneven floor eliminates the need for a _____.

_____ 5. When laying out the wall, a level line is drawn _____ up the wall to indicate the top of the base cabinets.

_____ 6. When installing cabinets, most installers prefer to mount the _____ units first.

_____ 7. The installation of wall cabinets is started in a _____.

_____ 8. If base cabinets are to be fitted to the floor, then their level layout line is measured from the _____ of the floor.

_____ 9. Countertops are covered with a thin, tough, high-pressure plastic laminate known as _____.

_____ 10. To prevent scratching the countertop when making the sink cutout, apply _____ to the base of the saber saw.

Name _____  Date _____

# CHAPTER 88 CUSTOM-MADE CABINETS AND COUNTERTOPS

## Multiple Choice

Write the letter for the correct answer on the line next to the number of the sentence.

_____ 1. The level line for the bottom of the wall cabinets is placed _____ above the floor.
   A. 35¼"
   B. 37"
   C. 48"
   D. 54"

_____ 2. When building wall cabinets, the first items fastened to the wall are small ¾" by 1" strips of wood called _____ .
   A. toeboards
   B. stile
   C. cleats
   D. mullions

_____ 3. The _____ the next component of a custom-built cabinet to be installed.
   A. cabinet ends are
   B. top rails are
   C. wall cabinet base is
   D. bottom rails are

_____ 4. From the top of the cleats, level lines are drawn on the inside faces of the cabinet end's cleats for the position of the _____ .
   A. countertop
   B. backsplash
   C. shelves
   D. face frame

_____ 5. In cabinets over 4' wide _____ must be installed for support between the cabinet top and bottom.
   A. headers
   B. partitions
   C. spreaders
   D. parting strips

_____ 6. Horizontal members of the face frame are called _____ .
   A. stiles
   B. rails
   C. trimmers
   D. preachers

245

_____ 7. Intermediate vertical members of the face frame that divide the total cabinet width into door openings are called _____ .
A. mullions
B. jambs
C. muntins
D. tenons

_____ 8. The bottom of the base cabinet is supported against the wall by a cleat whose top edge is installed to a level line drawn _____ from the floor.
A. 2"
B. 2½"
C. 3¼"
D. 4"

_____ 9. The inside of the toe board is located _____ inches away from the wall.
A. 18"
B. 20¾"
C. 22⅛"
D. 24"

_____ 10. The cabinet bottom must fit the wall with its outside edge overhanging the toe board by _____ from one end of the cabinet run to the other.
A. 1¾"
B. 2¼"
C. 2¾"
D. 3¼"

## Completion

**Complete each sentence by inserting the correct answer on the line near the number.**

_____ 1. The countertop drawer slides and shelves are supported against the wall by horizontal _____ .

_____ 2. Cabinet ends should be cut to a rough length of about _____ inches.

_____ 3. The rough width of cabinet ends should be about _____ inches.

_____ 4. Most countertops are covered with _____ .

_____ 5. Before laminating a countertop, lightly hand or power sand all joints making sure they are _____ .

_____ 6. When trimming laminate with a router or a laminate trimmer, it is recommended to use a _____ trimming bit.

_____ 7. _____ cement must be dry before the laminate is bonded to the core.

_____ 8. To test the cement for dryness you can check it with your _____ .

_____ 9. If the cement is allowed to dry more than ____ hours, the laminate will not bond properly.

_____ 10. When using a trimming bit with a dead pilot, the laminate must be _____ where the pilot will ride.

_____ 11. To prevent water from seeping between the backsplash and the countertop, apply _____ to the joint.

_____ 12. If heated to 325°F, the laminate can be bent to a minimum radius of _____ inches.

## Discussion

**Write your answer on the lines below the instruction.**

1. What are some of the reasons why it might be advantageous to custom build cabinets on the job, rather than purchase factory-built ones?

_____

_____

_____

**247**

Name _____  Date _____

# CHAPTER 89  DOOR TYPES, CONSTRUCTION, AND INSTALLATION

## Multiple Choice

Write the letter for the correct answer on the line next to the number of the sentence.

_____ 1. Designs may be routed into the faces of solid doors with _____ bits.
   A. straight
   B. V-groove
   C. core box
   D. A, B, and C

_____ 2. Strong joints can be made on the frame of a paneled door by using _____.
   A. contact cement
   B. a biscuit joiner
   C. small strips of metal lath for reinforcement
   D. expansion anchors

_____ 3. Grooves are cut on one edge of each piece of panel frame ¼" deep and as wide _____.
   A. as the thickness of the panel stock
   B. as ½ the thickness of the panel stock
   C. as twice the panel stock thickness
   D. ¾ the panel stock thickness

_____ 4. When assembling a paneled door, _____.
   A. glue the panel to the frame
   B. don't glue the frame joints
   C. rack the door square after the glue sets
   D. glue the frame joints but not the panel edges

_____ 5. Wrought iron and other decorative hinges are usually _____ hinges.
   A. pivot
   B. concealed offset
   C. European style
   D. surface

_____ 6. Offset hinges are used on _____ doors.
   A. overlay
   B. flush
   C. lipped
   D. European style

_____ 7. Pivot hinges are usually used on _____ doors.
   A. overlay
   B. flush
   C. lipped
   D. European style

_____ 8. Many carpenters use a self-centering tool, called a _____, when drilling pilot holes for a screw fastening of cabinet door hinges.
   A. VIX bit
   B. expansive bit
   C. butt gauge
   D. butt marker

_____ 9. When two screws are used to fasten a pull, drill the holes _____ the diameter of the screw.
   A. smaller than
   B. the same size as
   C. slightly oversized
   D. twice

## Completion

Complete each sentence by inserting the correct answer on the line near the number.

_____ 1. Cabinet doors are classified by their construction and also by their method of _____.

_____ 2. The most widely used method of hanging cabinet doors is the _____.

_____ 3. Face frames are not used on _____ style cabinets.

_____ 4. The _____ door has rabbeted edges that overlap the opening by about ⅜" on all sides.

_____ 5. The _____ type door must be fitted to the door opening.

_____ 6. Particleboard doors are usually covered with _____.

_____ 7. To give added clearance to the swing of a lipped door, cut the rabbet on a slight _____.

_____ 8. The pieces of a solid matched board door are held together by _____ that are attached with screws near the top and bottom of the door's back.

_____ 9. When laminating doors, both sides must be sealed to the same degree to prevent them from _____.

_____ 10. Veneer strips used to band doors may be made on the job by slicing thin strips with the _____.

250

Name _____  Date _____

# CHAPTER 90  DRAWER CONSTRUCTION AND INSTALLATION

## Completion

Complete each sentence by inserting the correct answer on the line near the number.

_____  1. Drawer fronts are generally made from the same material as the cabinet _____.

_____  2. Drawer sides and backs are usually ____ inch thick.

_____  3. Drawer bottoms are usually made of ____ inch thick plywood or particleboard.

_____  4. The _____ joint is the strongest used in drawer construction.

_____  5. The _____ joint is the easiest joint to make that is used between the drawer front and side.

_____  6. To provide added strength, the drawer back is usually set a ____ inch from the rear of the sides.

_____  7. On an overlay drawer, the overlay front is fastened to the _____ front with screws from the inside.

_____  8. The height of a drawer side should be about ____ inch less than the drawer opening.

_____  9. For a standard base cabinet, drawer side lengths are usually ____ inches long.

_____  10. A clearance of ____ inch is needed between the drawer opening and the drawer sides for most metal side guides.

_____  11. If the drawers are to slide on wood guides, generally ____ inch clearance is needed on each side.

_____  12. Both ends of a drawer false front must be _____ to fit into the dado on the drawer sides.

_____  13. If the drawer sides are ½″, then the groove for the drawer bottom should be about ____ inch deep.

_____  14. Both ends of a lipped drawer front are rabbeted to receive the _____.

_____  15. The simplest type of wood drawer guide is probably the _____.

251

## Identification: Wood Joints

Identify each term, and write the letter of the correct answer on the line next to each number.

_____ 1. dovetail joint

_____ 2. dado joint

_____ 3. dado and rabbet joint

_____ 4. butt joint

_____ 5. lock joint

_____ 6. rabbeted joint

A.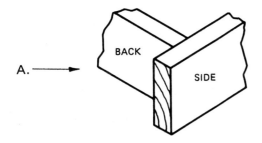

D.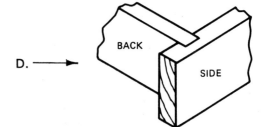

B.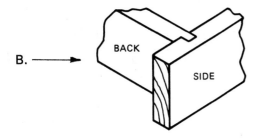

E.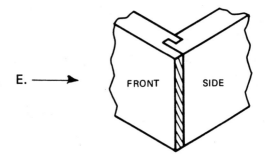

C.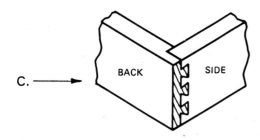

F.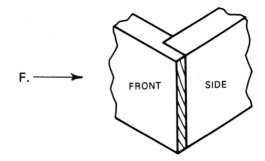